The Spiral Galaxy M33

Astrophysics and Space Science Library

For further volumes:
http://www.springer.com/series/5664

Paul Hodge

The Spiral Galaxy M33

Springer

Paul Hodge
Department of Astronomy
University of Washington
Seattle, Washington 98105-1580
USA
hodge@astro.washington.edu

Cover Figure caption: M33 in radio and optical.

ISSN 0067-0057
ISBN 978-94-007-2024-4 e-ISBN 978-94-007-2025-1
DOI 10.1007/978-94-007-2025-1
Springer Dordrecht Heidelberg London New York

Library of Congress Control Number: 2011938108

Printed on acid-free paper

Springer is part of Springer Science+Business Media (www.springer.com)

Preface

This book summarizes the gathering of knowledge of M33 from the 1920s, when Hubble first determined its true nature, to the twenty-first century, when the Hubble Telescope probed deeply into its many secrets. More than our Milky Way or the Andromeda Galaxy, the smaller spiral M33 is the only really typical spiral galaxy of our Local Group. It is neither a giant nor a dwarf. It has a regular symmetrical spiral structure, not conspicuously distorted by recent tidal encounters or interactions. While not face-on, it is not tilted too much (as is the case of M31) to make it difficult to untangle its structure. And it lies near enough to allow us to study its stars in enough detail to measure stellar and cluster ages and chemical composition, as well as to penetrate its complexes of gas and dust. In short, M33 is well-placed to be the object for the study of a typical spiral galaxy.

I have written this book with a structure that is modeled after that of my book of many years ago about its neighbor, M31 (Hodge P. 1992), but with a different aim in mind. When that earlier book was written, we did not have many powerful tools for retrieving astronomical literature. Then my material was gathered primarily by pouring over journal indices and collecting reprints and photocopies. At the time my aim was to provide a guide to this highly-dispersed literature for astronomers interested in gaining a start in the study of that important galaxy.

This new book is being brought into a new kind of fact-finding world. Because of ADS (the Smithsonian/NASA Astronomical Data System), virtually all astronomical literature is available right there on your laptop. Because of astro-ph (Cornell's astronomy-physics preprint service), most astronomers can read papers before they are even published. So the question now is, "What use is there for a book?" My answer has been to attempt to put together a manuscript that places current research on M33 (and similar galaxies) in perspective, both historical perspective and global perspective, and to do so in a language that can be appreciated by both specialists and non-specialists, by both professional astronomers and non-professionals, by both scientists and the curious public and, most importantly, by students.

I hope that this attempt has been successful and that the book will be either useful or interesting or both. As a person who knows a lot less than he would like to know, I have enjoyed learning what I've learned and thinking about what I haven't learned.

Finally, I want to express appreciation for the help that I have enjoyed. It is appropriate for me to acknowledge the excellent literature sources mentioned above and to thank the many scientists whose papers I have read. I am grateful to the astronomers who have allowed me to reproduce their figures and tables. And, last and most humbly, I am thankful to the publishers for their amazing patience throughout the years that it has taken me to complete this task.

Seattle, Washington, USA Paul Hodge

Reference

Hodge, P. 1992, The Andromeda Galaxy (Kluwer Academic publishers, Dordrecht)

Note on References

There are nearly 300 references to the astronomical literature in this book; they are listed at the end of each chapter. For the sake of efficiency, I have used the standard abbreviations for the names of journals that are commonly used in the astronomical literature, as listed below. For the more seldom-used journals, I have used more complete abbreviations.

Abbreviations to Common Journals' Names

AJ	Astronomical Journal
ApJ	Astrophysical Journal
ApJS	Astrophysical Journal Supplements
A&A	Astronomy and Astrophysics
A&AS	Astronomy and Astrophysics Supplements
Ap&SS	Astrophysics and Space Science
MNRAS	Monthly Notices of the Royal Astronomical Society
PASP	Publications of the Astronomical Society of the Pacific

Authors' Names

Because of the large numbers of authors found on astronomical papers in recent years, I have chosen to limit the number of names cited to three. Additional authors' names are indicated by the usual convention: "et al."

Copyright Notice

Contents

Chapter 1
Basic Data

There are three spiral galaxies in our Local Group: the Milky Way Galaxy (MWG), the Andromeda Galaxy (M31, NGC 224) and M33 (NGC 598). These three are the largest and brightest of the 50 or so members of the group. The others are all low-luminosity elliptical galaxies, irregular galaxies and dwarf spheroidal galaxies.

Compared to its two large spiral neighbors, M33 is a modest galaxy, rather typical of the majority of the spiral galaxies in our part of the universe. Because it is about average in size and luminosity and because its structure is relatively free of distortions caused by close neighbors, M33 is an important touchstone for the understanding of normal galaxies, especially late-type spirals.

1.1 Basic Data

The basic properties of M33 are now well-known, although some have remained in mild dispute. For reference, Table 1.1 provides the values of the data that will be used throughout this book. Many of these data follow those adopted in van den Bergh's splendid book, THE GALAXIES OF THE LOCAL GROUP (2000). Details of the other sources are given in the respective chapters in this book, where references to the literature are given.

1.2 The Distance to M33

The distance modulus given in Table 1.1 is representative of recent determinations by a variety of means and by different authors (Fig. 1.1). Table 1.2, which is expanded from the compilation made by van den Bergh (2000), provides a representative sample of distance determinations made in the last 50 years. The Cepheid

P. Hodge, *The Spiral Galaxy M33*, Astrophysics and Space Science Library 379,
DOI 10.1007/978-94-007-2025-1_1, © Springer Science+Business Media B.V. 2012

Table 1.1 Basic data for M33

Names	NGC 598, M33, Triangulum nebula	
Position (2000)	RA = 01 33 50.9, Dec = +30 30 37	de Vaucouleurs and Leach (1981)
Hubble type	Sc II–III	Sandage and Tammann (1981
Radial velocity (km/s)	−181	Deul and van der Hulst (1987)
Size (optical)	83′ × 53′	Holmberg (1958)
Size (HI)	135′ × 74′	Corbelli et al. (1989)
V	5.83	Jacobsson (1970)
B – V	0.6	"
U – B	0.00	"
E (B – V)	0.07	van den Bergh (2000)
A_V (foreground)	0.22	"
A_V (total)	0.40	"
A_B "	0.52	"
A_I	0.19	"
A_K	0.04	"
$(m - M)_0$	24.67	Table 1.2
Distance (kpc)	850 ± 20	"
M_V	−19.04	
ε	0.5	Sandage and Tammann (1981)
i	56°	Deul and van der Hulst (1987)
PA (major axis)	23°	de Vaucouleurs (1959)
Mass (optical extent; solar masses)	1.1×10^{10}	Deul and van der Hulst (1987)
HI Mass (radio extent; solar masses)	2.6×10^{9}	Corbelli and Schneider (1997)

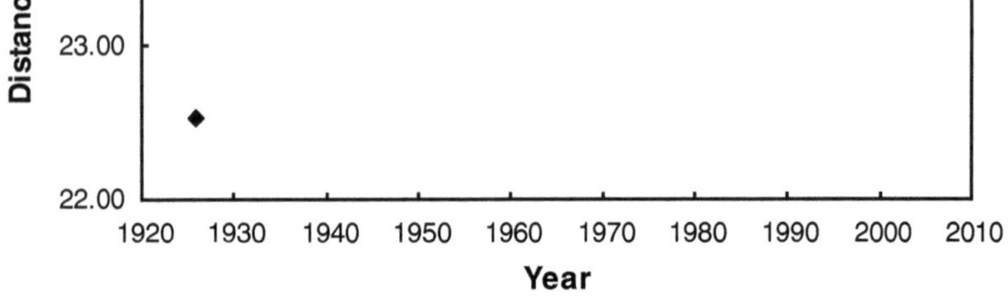

Fig. 1.1 Some values of the distance modulus for the galaxy M33 determined over the first 82 years of its study

Table 1.2 Distance measurements to M33

Reference	Method	$(m - M)_0$	Error
Hubble (1926)	Cepheids	22.53	
Sandage (1983)	Hubble (corr.)	24.83	
Sandage and Carlson	Cepheids	24.83	
Madore et al. (1985)	Cepheids	24.3	0.2
Freedman (1985)	Cepheids	24.1	0.2
Mould and Christian (1986)	Red giants	24.80	0.30
Kinman et al. (1987)	LP variables	24.55	
Christian and Schommer (1987)	Cepheids	24.2	0.15
Prichet (1988)	RR lyraes	24.60	0.23
Della Valle (1988)	Novae	24.38	0.10
Wilson et al. (1990)	Red giants	24.60	0.30
Mould et al. (1990)	LP variables	24.52	0.17
Freedman et al. (1991)	Cepheids	24.49	0.09
Greenhill et al. (1993)	Water masers	24.0	1.0
Sarajedini etal. (2000)	HB stars	24.84	0.16
Kim et al. (2002)	Red giants	24.81	0.1
Kim et al. (2002)	Red clump	24.80	0.05
Lee et al. (2002)	Cepheids	24.52	0.14
Brooks et al. (2004)	Red giants	24.72	0.14
Galleti et al. (2004)	Red giants	24.64	0.15
Argon et al. (2004)	Water maser	24.50	0.65
Scowcroft et al. (2009)	Metallicity corrected	24.53	0.11
Bonanos et al. (2006)	Eclipsing binary	24.92	0.12

distances are referred to the PL relation for the LMC, for which a reddening-corrected distance modulus of $(m - M) = 18.50$ is assumed here. The absorption values assumed are those given in Table 1.1.

The mean of the values in the table, excluding Hubble's 1926 distance, is 24.67, which corresponds to 850 kpc, which we adopt in this book. At this distance, an arc second equals 4.12 pc.

Chapter 2
Developing Knowledge

During the first two centuries of its recognition by humans, M33 was just a vague fuzzy object among the stars. In time it evolved from a bothersome curiosity to an intriguing and puzzling phenomenon. This chapter summarizes the slowly accelerating interest in it, starting with Messier's catalog and culminating in the early observations at the Mt. Wilson Observatory.

2.1 Early Visual Observations

Charles Messier is a name that all amateur astronomers know. He was an eighteenth century French astronomer whose interest and reason for fame was his ability to discover new comets. But that is not the reason why his name is familiar today. While comet hunting, Messier was continually being bothered by fuzzy objects that at first looked as if they might be comets, but which didn't move from night to night, so were permanent features in the sky. He kept track of their positions, eventually listing, in 1784, a catalog of 103 non-stellar objects. Interest in these seems to have been minimal; they weren't even curiosities for most astronomers. William Herschel, of course, was an exception. Number 33 in Messier's catalog was one of thousands of non-stellar objects that Herschel studied and tried to understand. In one of his papers (Herschel 1785), he listed M33 as one of the objects that might be reckoned as comparable to the Milky Way. Herschel's catalogs of non-stellar objects eventually were gathered together and supplemented by J. L. E. Dreyer (1888) into the New General Catalog, where M33 became NGC 598.

2.2 The Earl of Rosse and Spiral Structure

Herschel's leadership in the study of the heavens was partly the result of his building larger telescopes than anyone else. His giant tubes were not significantly surpassed until the Earl of Rosse built, in Ireland, a mammoth telescope with a 72-in. diameter

P. Hodge, *The Spiral Galaxy M33*, Astrophysics and Space Science Library 379,
DOI 10.1007/978-94-007-2025-1_2,

Fig. 2.1 Lord Rosse's giant telescope with which he observed M33

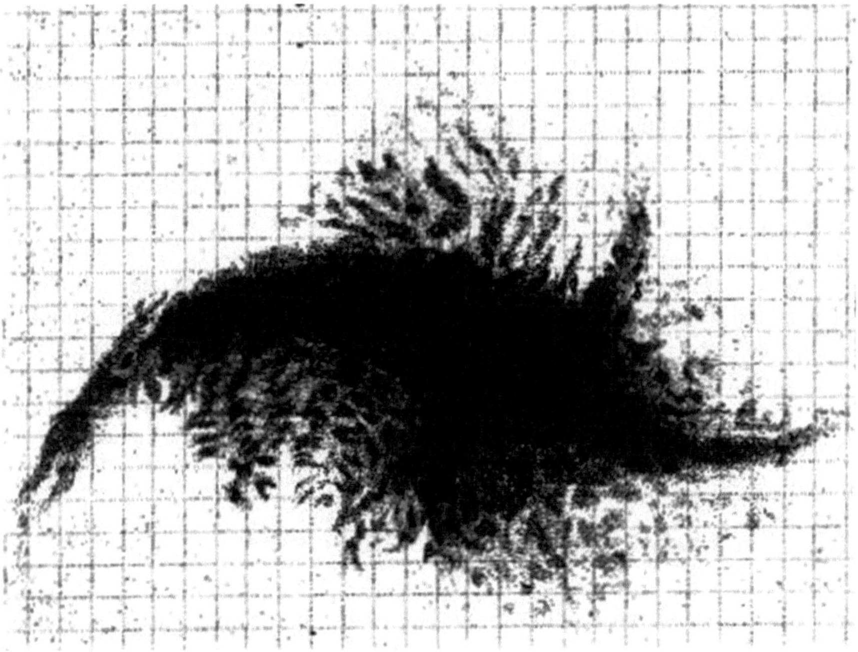

Fig. 2.2 The Earl of Rosse's sketch of the spiral structure of M33 (Parsons 1850)

mirror (Fig. 2.1), allowing unprecedented views of faint objects. He pointed the awkward giant at various nebulae and was the first to discern the spiral structure of M33 (Parsons 1850).

The central area was sketched as having a flocculent, diffuse spiral structure (Fig. 2.2) and the nebula was said to extend faintly "very far following".

In a later paper, he drew a more accurate representation of the spiral structure, with a prominent northern arm, a bright southeast arm and a stubby arm in the southwest. All of these components are recognized in modern photographs, as is nicely demonstrated in the excellent review of M33's history (to which this chapter is in debt) written by Kurtiss Gordon (Gordon 1969), where Rosse's drawing is compared to a Palomar Schmidt photograph (in Gordon's figure Rosse's drawing is rotated slightly clockwise with respect to the photo).

2.3 Roberts and the First Photographs

In the late nineteenth century astronomy was being transformed from a visual science based on drawings by eye and measurements by micrometer to a photographic science, with its advantages of permanence, precision and penetration to fainter levels. The first good images of M33 were taken by Isaac Roberts (1895), also in Ireland, who rigged up a camera to a telescope in his private observatory to record and study the nebulae. Roberts' photographs of M33 showed vastly more detail than was recorded visually and demonstrated both that the nebula was dominated by intricate spiral structure and was far larger than had been thought (Fig. 2.3).

A few years later in California, James Keeler (1908) was using the 36-in. Crossley Reflector to photograph nebulae and among these was M33.

2.4 Cataloging Condensations

While Roberts and Keeler were primarily interested in using their photographs for characterizing and classifying the various types of nebulae and star clusters, others with a more quantitative bent used them for more analytical purposes. For example, Dreyer (1904), working in Ireland, was especially interested in measuring condensations in nebulae to see whether any movement or other changes could be detected. Using the Roberts photographs, he measured positions of condensations in M33 and published a catalog of them.

Wolf and Ernst (1913) obtained photographs of M33 at the Heidelberg Observatory from which they measured positions of 517 distinct condensations. From these they were able to trace the structure to distances out far beyond the obvious extent of the nebula. They argued that something that subtends that large an apparent angle in the sky must be a relatively small distance away. Had they realized,

Fig. 2.3 A photographic image of M33 taken by Roberts (1895)

of course, that M33 was a stellar system like the Milky Way, they could have turned the argument around, arguing that it must be a large distance away. They stated that the condensations in their images fit onto the spiral arms of the nebula. Most of the cataloged condensations within the galaxy's image are now recognized to be the brightest OB associations and HII regions.

2.5 Fitting the Spiral Arms

An early example of making quantitative use of the catalogs was the paper by von der Pahlen (1911), who explored the mathematical characteristics of M33's arms, finding that each of the two prominent arms could be fit to logarithmic spirals, but with different pitch angles. Also he found that the arms were not exactly opposite each other, but were separated by several degrees in azimuth.

2.6 Radial Velocities and Rotation

In the early 1900s it became possible to make spectrographic observations of spiral nebulae, which produced some surprising results. Almost all showed large radial velocities, larger than for any other class of celestial objects, and most of them were negative.

Exceptions were the velocities of the two brightest spiral nebulae, M31 and M33. The first velocity for M33 was obtained by Pease (1915) using the 60-in. Mt. Wilson telescope. The great difficulty of this observation can be illustrated by reprinting his description as published in the PASP:

> 2. Messier 33 — The slit was set on the brightest knot in the nebula about 10 minutes *nf* the nucleus. The exposure was $34^{1/2}$ hours. The bright lines only appear, 5007, $H\beta$, $H\gamma$. The radial velocity is —278^{km}

The interesting fact about this is that the exposure was a whopping 34.5 h, meaning that he had to hand-guide the giant telescope, peering at a guiding star, for at least three nights. And all he was able to record on his tiny glass plate were three little images of emission lines. But he did get an answer and it was surprisingly large and negative. This value, ten times larger than typical stellar radial velocities, presented a puzzle as astronomers grappled with the problem of determining the distance to M33. With such a large radial velocity and such a large angular extent implying proximity, the proper motion should be detectable, but attempts to detect it mostly failed.

A further development was the discovery of what appeared to be rapid rotation of the spiral nebulae. Pease (1916) found that there was a radial velocity difference between the nucleus and an outer emission region (NGC 604) of 200 km/s, indicating probable rotational motion. This discovery was welcomed as a possible avenue for measuring the distances to the nebulae, because if measurements of the motions of condensations in the nebulae could be detected, then they could be combined with the observed radial velocity differences to obtain a measure of the distances. This idea led to the historically-unique phenomenon known now as "van Maanen rotation".

2.7 van Maanen Rotation

A. van Maanen was a Mt. Wilson astronomer who specialized in precise measurement of stellar positions. Despite suffering from nyctophobia (a remarkable disability for an astronomer), van Maanen was a productive and highly-respected scientist. It is unfortunate that today his name is known primarily in connection with a disastrous incident that was probably not his fault.

In the years following the installation of the Mt. Wilson 60-in. telescope in 1908, G.W. Ritchey obtained some splendid photographs of nebulae, showing more detail than ever before. It was realized that second epoch plates should be obtained so that any motion of individual condensations could be detected and then it might be possible to measure the rotation of the spiral nebulae and thereby obtain their distances. New plates were taken and by 1916 the first results were published. The measurements were made by van Maanen, the local expert in astrometry (van Maanen 1916 and several subsequent papers). In his measurements of hundreds of knots in several nebulae, he detected significant rotation and was able to combine his data with the available spectroscopic radial velocities to show that the spiral nebulae must be small, nearby objects.

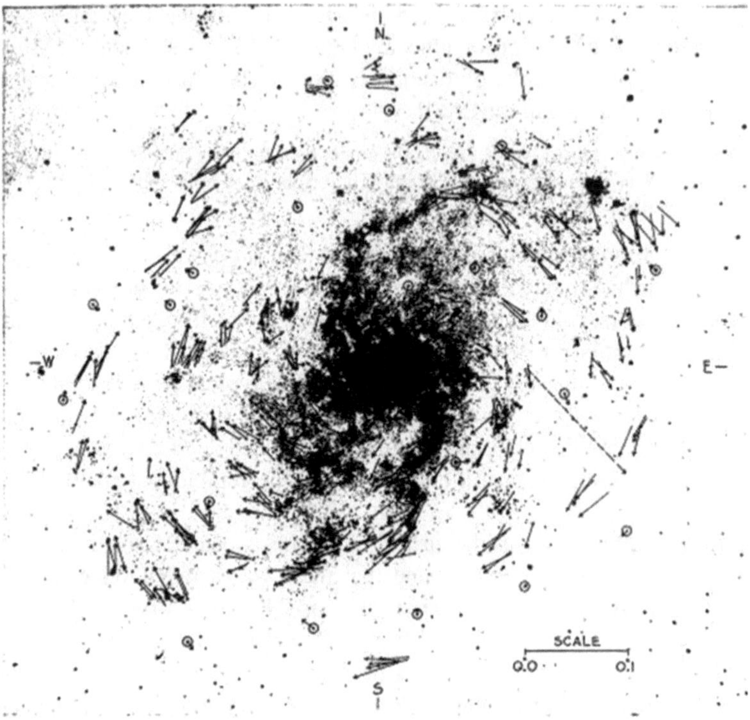

Fig. 2.4 van Maanen's map of motions in M33. The direction and length of the arrows indicate the amount of motion (van Maanen 1923). Used by permission, copyright AAS

In his results for M33 (van Maanen 1923) he claimed a period of rotation of 60,000–240,000 years. This clearly ruled out the idea, promoted in some circles, that M33 might be a remote external object beyond the edge of the Milky Way, as the implied velocities would be unbelievably enormous (Fig. 2.4).

Unfortunately, it was shown later that all of van Maanen's rotations were spurious. Measurements made on Mt. Wilson 100-in. telescope plates showed no detectable rotation (Hubble 1935; van Maanen 1935). Various theories have been put forward to explain what had gone wrong. One idea was that the spiral structure of the nebulae made them look so much like they were rotating that this subconsciously influenced van Maanen as he worked. That hypothesis has been used for decades as a warning to science students to be exceptionally careful to guard against such unconscious bias.

An alternate explanation was proposed by Walter Baade (1963), who knew that Ritchey's 60-in. plates owed their superior definition to the fact that he would expose plates only during times of excellent seeing. When the images that he was using for guiding showed moments of poor definition, he would swing the plateholder out of the focal plane and wait until the seeing improved again. Baade's belief was that this rotation of the plateholder led to slightly asymmetrical stellar images, which translated into an apparent rotation of the nebula.

Either explanation of "van Maanen rotation" makes a useful lesson for anyone trying to make measurements near the limit of any technique.

2.8 Lundmark's Studies

Knut Lundmark was a Swedish astronomer whose ideas and accomplishments have been unreasonably neglected by history. He did some advanced experiments and thinking and arrived at several important conclusions that were contemporaneous with the ideas of the better known Mt. Wilson astronomers.

Lundmark's interest in spiral nebulae led him to do some analysis of the structure of M33 (Lundmark 1921) and he devised an ingenious method for gauging its distance from star counts (Lundmark 1926). He counted stars to different limiting magnitudes in and around M33 and showed that the distribution of stars at M33's position was the same as in the surroundings down to blue magnitudes of 15.7, but fainter than that the numbers slowly increased. Arguing that if the brightest stars in M33 were as bright as the average brightness of stars near the sun, then M33 must be at least 3,000 pc distant. If, however, these stars were as bright as the brightest known stars, with absolute magnitudes of –6, then M33 must be 300 kpc distant, placing it far beyond the edge of the Milky Way. This distance, in fact, is almost the same as that derived by Hubble 5 years later from Cepheid variables (Chap. 3).

As a footnote to the van Maanen rotation story it should be mentioned that Lundmark was sufficiently concerned about the discrepancy between van Maanen's result and the growing evidence that favored a large distance to the spiral nebulae that he completely remeasured van Maanen's plates in 1926. He derived motions that were only about 10% as large as van Maanen's, though they suggested a rotation in the same direction (Lundmark 1926). This result seems to confirm Baade's explanation, for if the images had spurious rotational asymmetry, the amount of displacement would be expected to vary with the way the measurer will have centered the crosshairs on the images.

2.9 The First Variables

One of the early hints that spiral galaxies might be large, distant systems of stars was the discovery of apparently stellar images that varied in brightness. It was a major step forward when Duncan reported the discovery of three periodic variable stars in M33, as well as a possible nova (Duncan 1922). The variables were discovered with the Mt. Wilson 60-in. telescope and Duncan then followed up by measuring them on plates taken earlier with the telescopes at the Mt. Wilson, Lick, Lowell and Yerkes Observatories. Their variability was conspicuous, but Duncan was unable to measure a period, such as would indicate that they might be Cepheid or

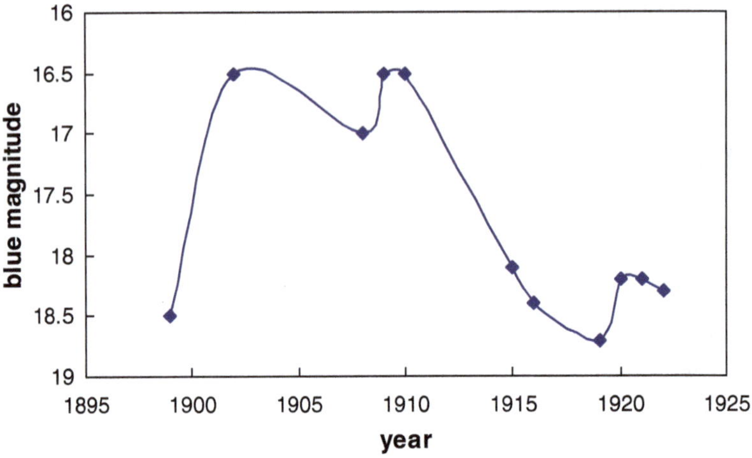

Fig. 2.5 The light curve of M33's variable #1 as measured by Duncan (1922). The curve shows no regularity

long period variables (Fig. 2.5). Four years later Hubble examined them on his new set of plates and called them irregular variables (later Hubble-Sandage variables or LBVs, see Chap. 10).

2.10 Summing Up

After being noticed and cataloged by astronomers including Charles Messier and William Herschel, M33 remained largely unstudied for more than a century. Visual observations and, later, photographs showed its pinwheel shape, similar to that of other cloud-like nebulae with spiral structure, but its properties remained mysterious. Only when the large telescopes of the twentieth century turned towards it were clues discovered that eventually led to understanding its true nature.

References

Baade, W.: Evolution of Stars and Galaxies. Harvard U. Press, Cambridge (1963)
Dreyer, L.E.: MemRAS **49**, 1 (1888)
Dreyer, L.E.: Proc. Roy. Irish Acad **25A**, 3 (1904)
Duncan, J.C.: PASP **34**, 290 (1922)
Gordon, K.: QJRAS **10**, 293 (1969)
Herschel, W.: Phil. Trans. Roy. Soc. Lon. **75**, 213 (1785)
Hubble, E.P.: ApJ **81**, 334 (1935)
Keeler, J.E.: Pub. Lick. Obs. **8**, 1 (1908)

Lundmark, K.: PASP **33**, 324 (1921)
Lundmark, K.: ApJ **63**, 67 (1926)
Parsons, W.: Phil. Trans. Roy. Soc. **140**, 499 (1850)
Pease, F.G.: PASP **27**, 239 (1915)
Pease, F.G.: PASP **28**, 33 (1916)
Roberts, I.: MNRAS **56**, 70 (1895)
van Maanen, A.: ApJ **44**, 210 (1916)
van Maanen, A.: ApJ **57**, 264 (1923)
van Maanen, A.: ApJ **81**, 336 (1935)
von der Pahlen, W.: AN **188**, 249 (1911)
Wolf, M., Ernst, E.: Veroff. Der Bad. Sternwarte zu Heidelberg **6**, 115 (1913)

Chapter 3
Hubble's Pioneering Study

In 1924 Edwin Hubble published his first paper demonstrating the extragalactic nature of a "nebula", the dwarf irregular galaxy NGC 6822 (Hubble 1925). A year later he presented a long and exacting paper on M33 that clearly demonstrated that it is also a distant stellar system (Hubble 1926). A similar paper, better known to history, is his massive study of M31, published 3 years later (Hubble 1929). Because the M33 paper laid the groundwork for many of the studies of that galaxy (and others) that followed, it is interesting to review it in detail. Note that all of the figures in this chapter are reproduced from his paper in the Astrophysical Journal and are reproduced with permission from the American Astronomical Society.

3.1 The "Condensations" Are Stars

For decades before Hubble's work, the spiral nebulae M33 and M31 were said to contain "condensations". The nature of the compact images was uncertain. Some considered them to be "nebulous stars" (Ritchey 1910), while others thought that they might be ordinary stars (e.g. Lundmark 1921). Spectroscopy of the brightest of them, obtained by Lundmark (1925), showed a continuous stellar-like spectrum. Hubble criticized the "nebulous star" terminology by pointing out that most photographs, especially the excellent Ritchey plates, were exposed with the nebula's nucleus in the center; this meant that the inner condensations were imbedded in the nebulous body of M33's center, making them appear nebulous. Also the outer objects were in the areas on the plates where aberrations distorted the images, so that they also did not appear starlike. At the same time, he criticized Lundmark's conclusion by insisting that the spectra were inconclusive as to the nature of the condensations.

Hubble settled the question by obtaining a series of plates of M33, some exposed towards the nucleus and others centered on the outer parts (Fig. 3.1), during some exceptionally good seeing at the Mt. Wilson 100-in. telescope. He found that the condensations were identical in size to the images of stars in a nearby Selected

P. Hodge, *The Spiral Galaxy M33*, Astrophysics and Space Science Library 379,
DOI 10.1007/978-94-007-2025-1_3, © Springer Science+Business Media B.V. 2012

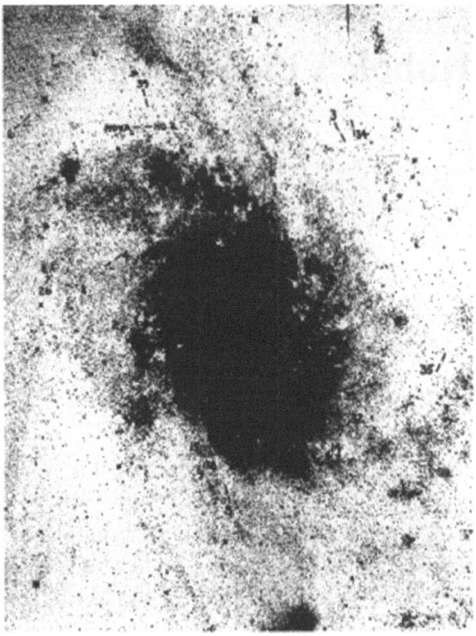

Fig. 3.1 Hubble's published plate of M33 taken with the Mt. Wilson 60-in. telescope. Used by permission, copyright AAS

Area, measured to be 0.55 arcsec on the best plates (the 100-in. had then and continues to have occasional nights with superb seeing). His conclusion was that the "nebulous stars" were, in fact, normal stars. He suggested that their faintness alone was evidence of the great distance to M33, as star groupings with faint dwarfs as their brightest stars were unknown.

3.2 M33's Variable Stars

Before Hubble's work, Duncan (1922) and Wolf (1923) independently discovered variable stars in the direction of M33. Encouraged by this, Hubble used both existing 60-in. plates and new plates that he obtained with the 100-in. telescope, gathering together a collection of 80 photographs of M33. From a blink microscope survey, he identified 45 variable stars (Fig. 3.2). These he claimed to belong to M33, as the density of variables in the area covered was vastly greater than in the surrounding sky.

Hubble carried out photometry of the variables by first establishing a set of comparison stars for each variable star. He calibrated the comparison stars' magnitudes by means of photographic transfers from nearby Selected Areas, for which Seares had established magnitudes down to 18.5 photographic (blue). With no fainter magnitudes available, Hubble was forced to extrapolate his magnitude system further, claiming reliability only to 19.2, though he published measures as faint as 19.9

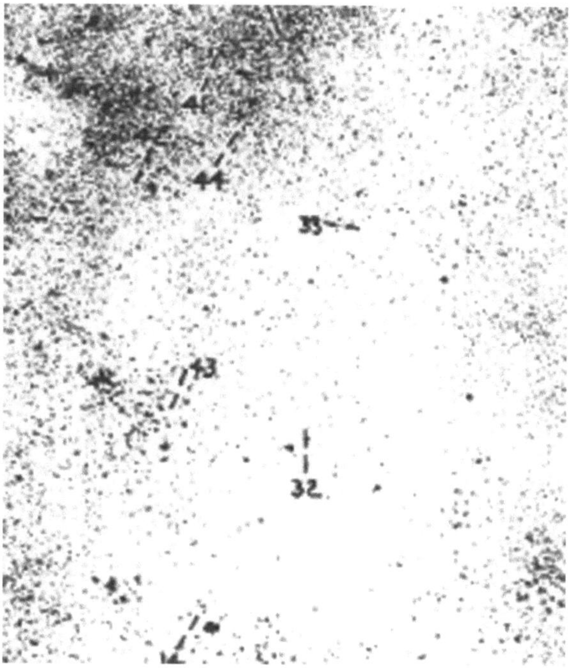

Fig. 3.2 Variables marked on one of Hubble's images of M33 (Hubble 1926). Used by permission, copyright AAS

Hubble found that his systematic errors were about 0.4 magnitudes, with individual measurements of the Cepheids at maximum being more reliable, within his magnitude system. All measurements were made by eye to the nearest tenth of a magnitude. [Many years later Sandage investigated Hubble's comparison star magnitudes in M33 and rederived Hubble's period-magnitude relation to obtain a new distance to the galaxy (Sandage 1983). He found that the Hubble magnitudes were systematically too bright at the faint limit of his measurements and that using corrected magnitudes and a modern PL relation, the derived distance to M33 was three times greater than Hubble's original distance.]

The result was the discovery that 40 of the variables could be classified as examples of known types of variable stars, the remaining five having too few observations to determine their nature. There were 35 Cepheids, one eclipsing star and four irregular variables.

3.3 The Cepheids

The Cepheids turned out to be excellent examples of that type of star, essentially identical to those found in the Milky Way. The range in period was 13–70 days and the amplitudes ranged up to 1.2 magnitudes. Light curves (Fig. 3.3) showed the characteristic Cepheid shape: a rapid rise in brightness followed by a slower decline.

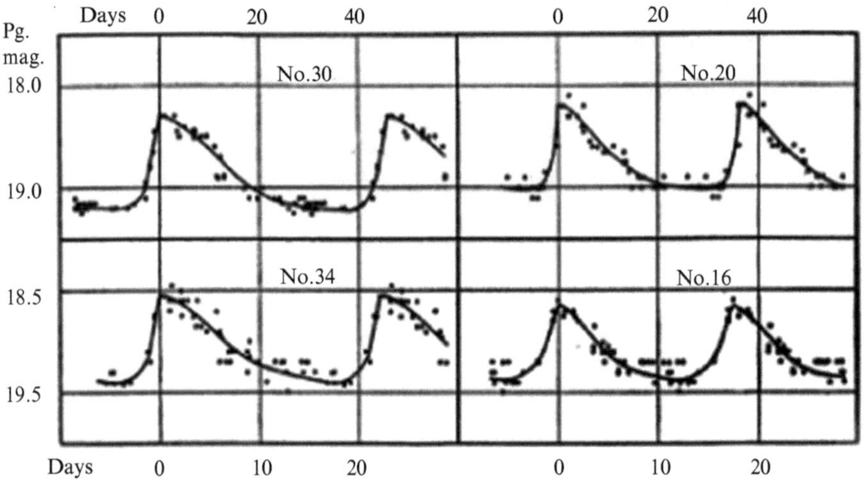

Fig. 3.3 Light curves for four Cepheids in M33 (Hubble 1926). Used by permission, copyright AAS

Although red-sensitive plates required much longer exposures at that time, Hubble was able to determine that the Cepheids had intermediate colors (like the sun's) and this further confirmed their identity as normal Cepheids.

Having established that his stars were Cepheids, Hubble then plotted their maximum magnitudes against the logarithm of their periods and obtained a period-magnitude relation similar to that of the Cepheids in the Small Magellanic Cloud. Using Harlow Shapley's latest version of that (Shapley 1925) and using a simple process of superimposing the diagrams and fitting them by eye, he concluded that the Cepheids in the two galaxies were similar and that the relative distances could be derived by the comparison. Using Shapley's distance modulus for the Small Magellanic Cloud, $(m - M) = 17.55$, Hubble derived a distance modulus for M33 of $(m - M) = 22.1$. This implied a distance of 263 kpc (Fig. 3.4).

This result obviously established M33 as a separate object, not a part of the Milky Way Galaxy. [Hubble did not use the term "galaxy" for external objects, preferring to refer to them as "extragalactic nebulae"]. He estimated the uncertainty in the distance to M33 as being about 10%. He was careful to mention the assumptions that were made in establishing this distance and was sensitive to the uncertainties in the magnitudes and to the possibility that the Cepheids in the two systems might not be the same and that the distance to the Small Magellanic Cloud might not be accurately known. But these concerns were dealt with by being mentioned and then dismissed. Time has shown that all of these concerns were warranted and that his estimate of the uncertainty in the distance to M33 was in error.

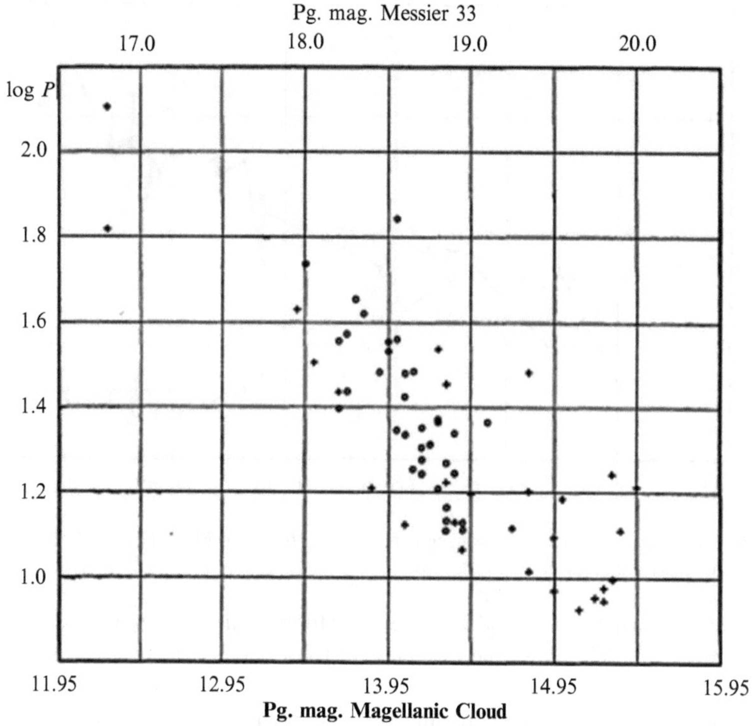

Fig. 3.4 Hubble's comparison of the period-luminosity relations in M33 (*circles*) and the Small Magellanic Cloud (*crosses*) (Hubble 1926). Used by permission, copyright AAS

3.4 Other Variables

Among the brightest variables were four stars that showed highly irregular variability. Two of them, numbers 1 and 2 in Hubble's list, were so bright that they could be measured on a variety of plates, including some of Roberts' taken as early as 1895. Their light curves are shown in Fig. 3.5.

Variable 2 became so bright in 1924 that it was the brightest star that Hubble felt could be assigned to M33. He obtained a spectrum of it with the 100-in. and characterized it as having a blue continuum with superimposed emission lines of hydrogen. The nature of these stars was barely discussed by Hubble, who limited himself to stating that "variation of this type is not sufficiently known among galactic stars for analogies to be traced".

[Analogies were traced later, however, by Hubble, working with his young disciple, Allan Sandage, who gathered data on these stars in a variety of galaxies and discussed their properties in terms of possibly using them as distance indicators.

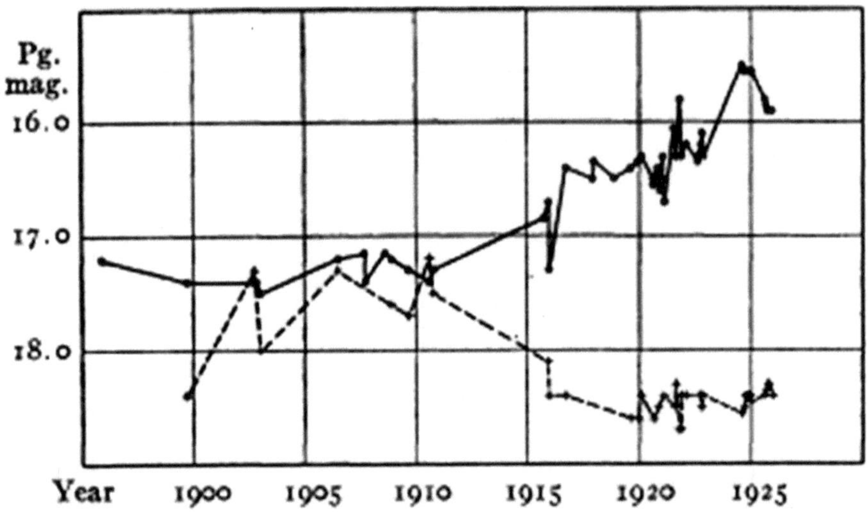

Fig. 3.5 Light curves for two of the irregular variables found in M33 (Hubble 1926). Used by permission, copyright AAS

They were known for a period of time as "Hubble-Sandage variables", but later came to be called "luminous blue variables" or LBVs; their current status is discussed in Chap. 10].

One star appeared to be an eclipsing variable, but its period was uncertain and its light curve incomplete. Its amplitude was approximately 0.6 mag.

3.5 Novae

Hubble observed two novae in M33, more than an order of magnitude fewer than he found in M31 during the same interval. The data were sparse, covering only a portion of the light curve of each. In one case only four data points were available, two of which were probably near maximum, while in the other case the object appeared to have been first seen when considerably past its maximum. Figure 3.6 shows their complete observations.

Though there was little coverage of the novae brightness, Hubble was able to say that, when compared to those in M31, the two novae had characteristics consistent with their being in a distant, separate stellar system at the distance given by the Cepheids.

3.6 The Diffuse Nebulae

Among the brightest objects seen in M33 are a number of diffuse features, some bright enough to have been separately cataloged in the NGC. Hubble particularly cited three, NGC 588, NGC 595 and NGC 604. He concentrated on the question of

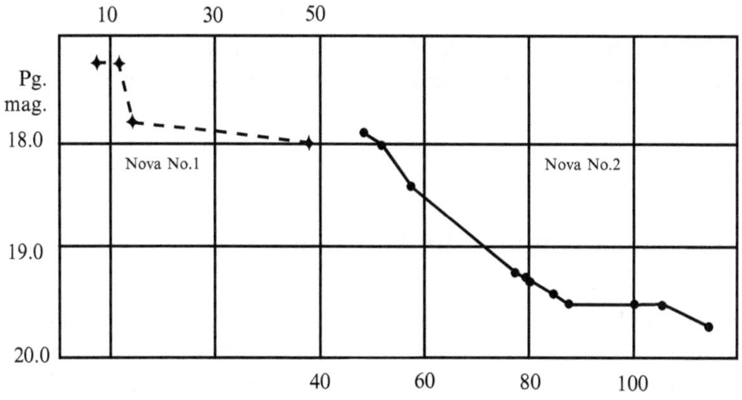

Fig. 3.6 Light curves of the two novae found in M33 (Hubble 1926). Used by permission, copyright AAS

whether or not these features were the same kind of objects as those identified as diffuse nebulae (now more commonly referred to as HII regions) in our Galaxy.

First, he pointed out that they had emission spectra, although it is not clear from his paper whether he took spectra of them or was basing his statement on others' observations. Furthermore, he stated that a slit spectrogram of NGC 604, the brightest, showed that the H-beta line of hydrogen was brighter than [NII], similar to what is common in Galactic HII regions.

He then turned to the stars detected in the diffuse nebulae. From red-sensitive plates, he found that the brightest stars were very blue, as is the case for Galactic HII regions. The apparent magnitudes of the brightest stars ranged from about 15.75 for NGC 604 to 16.6 for NGC 595. Assuming the Cepheid distance to M33, Hubble calculated that the brightest stars would have absolute magnitudes of about –6. There were luminous giants known in our Galaxy with absolute magnitudes of this order, though the average given for O giants was –4.4 (Plaskett 1924). Assuming Plaskett's estimate of about 3 magnitudes for the dispersion in the luminosities of O giants, Hubble concluded that the brightest stars in M33's nebulae were on the same order of absolute magnitude as those in our galactic system, if the Cepheid distance is correct.

In anticipation of their use later in the century as distance indicators, Hubble next discussed the sizes of the diffuse nebulae. The three brightest were found to have mean dimensions of 35, 55 and 75 parsecs, respectively, on the basis of the Cepheid distance. By comparison, he stated that the Orion loop in our Galaxy is 40 pc in diameter and NGC 2070 (30 Doradus) in the Large Magellanic Cloud is 80 pc in size. Thus he argued that the M33 objects are of a scale similar to that of the largest diffuse nebulae in other stellar systems.

Hubble went a step farther. Based on experiments that he carried out for our Galaxy's diffuse nebulae, he found a relationship between the apparent magnitude of the associated stars or group of stars, m, and the angular size of the nebula, a, given as

$$m + 5\log a = 11.1 +/- 0.8$$

Using integrated (blue) magnitudes for M33 and the sizes measured on his deepest plates, Hubble found a mean value of 11.8 for the sum of $m + 5\log a$, a value within the galaxy's dispersion, supporting the idea that these objects were similar in the two galactic systems.

3.7 The Stellar Luminosity Function

At the time of Hubble's work on M33 there were several astronomers involved in exploring our Galactic system by means of star counts and statistical distances. In the year of his M33 paper, Hubble was able to use a new study of the Galactic stellar magnitude distribution just completed by F. H. Seares and P. J. van Rhijn (1925). Therefore he was able to measure the luminosity function for M33 stars, subtracting the foreground Galactic stars using the Seares and van Rhijn tables.

Hubble counted stars in three fields across the face of M33, totaling 2534 stars, including foreground stars. After subtracting the foreground, the result was a stellar luminosity function that could be compared to that for the stars near the sun as determined by Kapteyn (1922). The two curves turned out to be remarkably similar, again indicating that M33 is a distant stellar system with a stellar population like that of our system.

Using his estimate that the star counts represented about 0.2 of the total area of M33, Hubble calculated a comparison between M33's volume and that of the Kapteyn tables for the solar neighborhood. Then, assuming that the M33 luminosity function (Fig. 3.7) continued to agree with the Kepteyn relation to fainter magnitudes, he was able to calculate the expected total apparent magnitude of M33 itself, obtaining a value of 7.1. This compared well with current measurements, which gave 7.0. Hubble argued that M33 must have a stellar luminosity function that agrees with our system's down to much fainter magnitudes than the three brightest magnitudes that he had measured. Thus M33 must have a total stellar makeup very much like that of our Galaxy.

3.8 The Nucleus

Although it occupies a very small portion of his paper, Hubble does discuss the nature of the nucleus of M33. He points out that it is probably not stellar, as his best short exposures indicated that it was resolved and appeared as a "globular nebula". Were it a star, he pointed out that it would be anomalously bright, with an absolute visual magnitude of −9, comparable to the strange star S Doradus in the Large Magellanic Cloud.

Instead Hubble concluded that it is similar to the non-stellar nuclei of other spiral nebulae, which have continuous spectra of fairly late type (e.g. he gave type G5 for

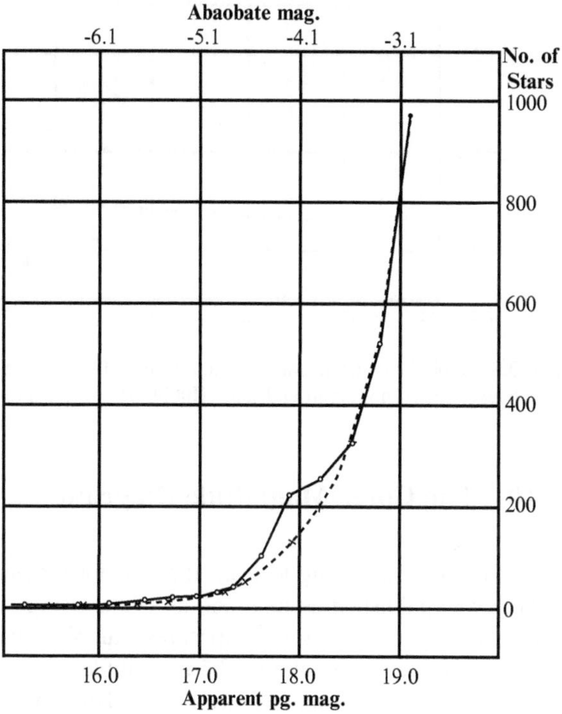

Fig. 3.7 The luminosity function for stars in M33 (*open circles*) and in our local galactic system (*crosses*) (Hubble 1926). Used by permission, copyright AAS

M33's nucleus). In a footnote he stated that two examples of spiral nebulae with unresolved nuclei were NGC 4051 and NGC 4151 and said that the spectra of these nuclei were "planetary", meaning similar to the emission line spectra of planetary nebulae. Twenty years later Carl Seyfert was to study these and other similar galaxies to reveal their remarkable nature; they are now called Seyfert galaxies and were the first examples of galaxies with active nuclei to be recognized.

3.9 The Mass

Having established that M33 is a separate, distant system of stars, Hubble used a simple argument to derive its mass. The radial velocity of the nucleus had been determined spectroscopically as −70 km/s and that of NGC 604, about 700 arc seconds northeast of the nucleus near the major axis, had been measured at −270 km/s. This gave a difference of about 200 km/s. Correcting for the tilt of the plane of the system, estimated as 42°, Hubble showed that the rotational velocity in the plane at the distance of NGC 604 must be about 270 km/s, implying a mass interior to that position of 1.5×10^{10} solar masses. Considering all of the assumptions and errors in the distance, this is a remarkably good estimate, comparing well with the present value for the optically visible galaxy of 1.1×10^{10} solar masses.

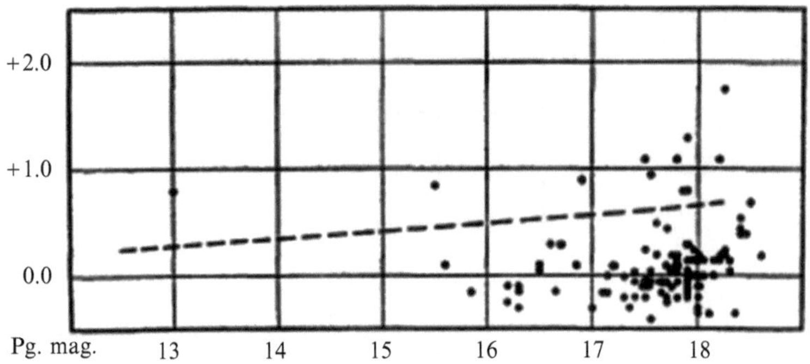

Fig. 3.8 Hubble's color-magnitude diagram for M33.The *dashed line* indicates the limit below which he expected no foreground stars (Hubble 1926). Used by permission, copyright AAS

3.10 The Color-Magnitude Diagram

It might seem surprising to learn that a color-magnitude diagram for an external galaxy was published as long ago as 1926. Using panchromatic plates, a yellow filter and long (as long as 3 h) exposures, Hubble obtained visual-wavelength plates of M33 that reached to about 18.5 magnitude. He measured (by eye) 28 stars that appeared on both yellow and blue plates and determined both photographic (pg) and photovisual (pv) magnitudes (using, it must be mentioned, a highly round-about way to calibrate the visual magnitudes). The result was a very nice color-magnitude diagram (Fig. 3.8), which looks very modern, except that he plotted it with the axes rotated 90° with respect to the contemporary custom.

He concluded that the brightest stars in M33 must almost all be blue. One of the redder stars was a variable and he suggested that the red side of the color-magnitude diagram might be occupied only by variable stars.

3.11 Summing Up: M33 and Hubble in Perspective

From the point of view of astronomy's fast-paced and highly-competitive nature in the twenty-first century, it is interesting to note two aspects of Hubble's work on M33 in the early twentieth century. First, he waited to publish his results all together in one long paper, covering all aspects of his exploration of the object. This one paper included the important implications of the results, which, together with his other two papers of the time (on M31 and NGC 6822), revolutionized our concept of the universe. Second, for years he had essentially no competition, largely because he was given exclusive rights to work on these objects with the world's only telescope large enough to do the job effectively. The culture at Mt. Wilson at the time was to have different fields divided among the staff members so that no-one encroached upon another's area of research. Thus, with guaranteed access to the world's largest telescope, with no need to apply for research funds and with no danger of competition from other astronomers in his institution or elsewhere in the world, Hubble could pursue his research on M33 deliberately and carefully.

The basic theme of Hubble's 1926 paper was that M33 must be an extragalactic stellar system. He demonstrated this fact by establishing its large distance and by several arguments based on its content. His arguments were as follows:

1. The resolved objects in M33 were stars, not "nebulous knots", as previous described.
2. M33 contained Cepheid variables, which showed a period-luminosity relation that indicated a distance of several hundred kiloparsecs.
3. Novae were detected and their brightnesses suggested a large distance that agreed with the Cepheid distance.
4. Among the stars were some diffuse nebulae that had emission-line spectra. Their intrinsic dimensions, if at the Cepheid distance, were comparable to those of the largest emission nebulae in the Milky Way galaxy.
5. The luminosity function of the stars was similar in shape to the van Rhijn function and extrapolations to faint stellar magnitudes provided a total magnitude for M33 in agreement with its measured total magnitude.
6. Radial velocities showed a large difference between the center and outer parts of M33 and, if interpreted as rotation, implied a very large mass.
7. The color-magnitude diagram of the brightest stars, if the Cepheid distance is assumed, was similar to what was known about the Milky Way's brightest stars (actually, Hubble was learning more about M33's brightest stars than was known then about our system's most luminous stars).

References

Duncan, J.C.: PASP **34**, 290 (1922)
Hubble, E.P.: ApJ **62**, 409 (1925)
Hubble, E.P.: ApJ **63**, 236 (1926)
Hubble, E.P.: ApJ **69**, 103 (1929)
Kapteyn, J.C.: ApJ **55**, 242 (1922)
Lundmark, K.: PASP **33**, 324 (1921)
Lundmark, K.: MNRAS **85**, 890 (1925)
Plaskett, J.: Pub. Dom. Ap. Obs. **2**, 287 (1924)
Ritchey, G.W.: ApJ **32**, 26 (1910)
Sandage, A.R.: AJ **88**, 1108 (1983)
Sears, F.H., van Rhijn, P.J.: Proc. Nat. Acad. Sci. **11**, 358 (1925)
Shapley, H.: Harvard Circular, No. 280 (1925)
Wolf, M.: Ast. Nachr. **217**, 476 (1923)

Chapter 4
Structure

The structure of M33 at optical wavelengths has been a topic of interest for over 70 years. Because of its small distance and its moderate angle of inclination, M33 provides a good source of information on the detailed structure of a common type of spiral galaxy.

4.1 Visual Wavelength Classification

Hubble, of course, was the first to assign a Hubble type to M33. He considered it to be a typical example of a Sc spiral, with open, loose arms and a small central bulge (Hubble 1936). Subsequent elaboration of the Hubble scheme by de Vaucouleurs (1963) led to a classification of SAc(s), indicating that it has no bar and that the arms begin at the center. Sandage (1961) provided an illuminating discussion of its visual structure and classified it as Sc. Its luminosity class (van den Bergh 1960) is II–III.

Most current papers use the simple Hubble classification of Sc. Persons interested in using visual impressions for statistical purposes might use the more elaborate choice of SAc(s)II–III.

4.2 Optical Photometry

M33 was one of the first galaxies to have its structure measured electronically. Patterson (1940) produced an early map, based on analysis of microphotometer scans of a photographic plate of M33. This accomplishment was more a demonstration of a technique than an analysis of the galaxy, as the paper did not pursue the scientific conclusions possible from the map. Seyfert (1940) used photometric scans to measure the brightness and color distribution of images of M33 and other spiral nebulae. He found that the condensations in the spiral arms were bluer than the nuclear areas of the galaxies.

P. Hodge, *The Spiral Galaxy M33*, Astrophysics and Space Science Library 379,
DOI 10.1007/978-94-007-2025-1_4,

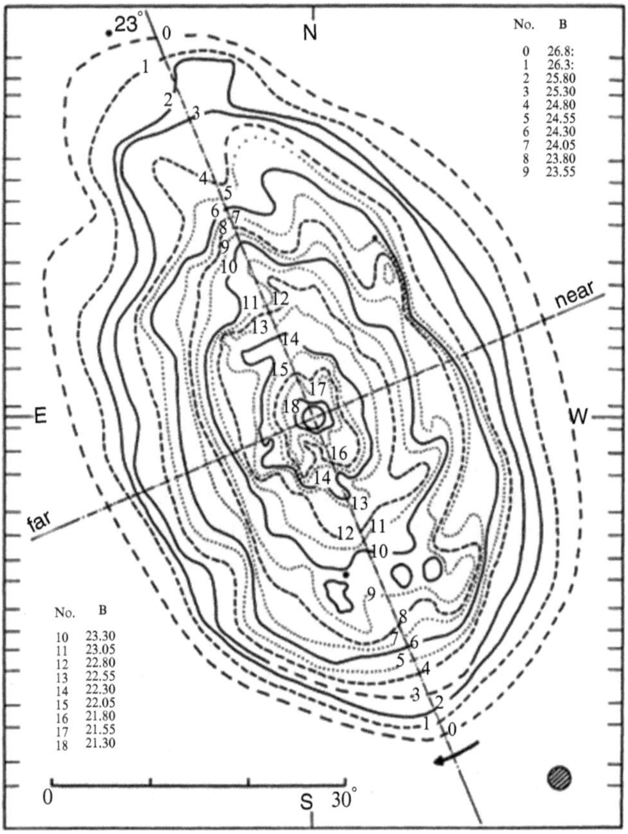

Fig. 4.1 An isophotal map of M33 obtained by de Vaucouleurs (1959). The isophotes are given in units of B mag/arcsec^2. The size of the scanning aperture is shown. Used by permission, copyright AAS

A more complete photometry of M33 was that of de Vaucouleurs (1959), who obtained diurnal photoelectric scans of the galaxy with a 53-cm telescope at Lowell Observatory (Fig. 4.1). Using an aperture of 2.3 arcmin and a spacing of 5 arcmin, the photometry covered the main body of the galaxy out to a surface brightness of approximately B = 26.8 mag/arcsec^2. The majority of the scans were in B, but there was also some photometry in V and U. These data provided a quite complete set of parameters for the galaxy: integrated magnitudes and colors, luminosity and color profiles, the axial ratio, the inclination angle, and the position angle of the major axis.

4.3 Integrated Properties

The apparent magnitude of a diffuse object of large angular size was a challenge for early visual observers. Hubble used a value for the apparent visual magnitude of 7.0 for M33. Shapley and Ames (1932) cataloged its magnitude as 7.8. A careful photographic photometry was published by Holmberg (1958). A photoelectric integrated measurement was made by de Vaucouleurs (1959), whose photometry gave a measured magnitude to the B=25.8 mag/arcsec^2 isophote of 6.30 +/– 0.02. He derived an asymptotic B magnitude of 6.27. Integrated colors were B – V=+0.55, U – B=–0.10. Thus the visual magnitude would be 5.72, considerably brighter than the Hubble or Shapley/Ames values.

4.4 Orientation

The orientation of the disk of M33 has been determined in many ways and the results are not in very satisfactory agreement. A surprising number of authors refer to M33 as being "nearly face-on," basing that on the general visual appearance of its open spiral structure. Actual measurements, however, show that its plane is tilted at a significant angle to the plane of the sky. For example, Danver (1942), using photographic plates and forcing the spiral arms to assume a logarithmic spiral form, derived a value for the inclination angle (with respect to the plane of the sky) of $i=40.4°$ with a position angle of PA=48.9°.

Using photoelectric photometry, de Vaucouleurs (1959) determined the axial ratio to be b/a=0.615 +/– 0.01. Assuming a thin plane, this corresponds to an angle to the line-of-sight of 35 +/– 1°, or to the plane of the sky of $i=55°$ if the intrinsic axial ratio of the spheroidal disk is q=0.1. The position angle was determined from the isophotes to be 23° (Fig. 4.2).

Boulesteix (1979) analyzed the distribution of the HII regions detected in their large survey, deriving values of $i=33°$, PA=20°. These values agreed satisfactorily with those of de Vaucouleurs (1959).

On the other hand, Sandage and Humphreys (1980) analyzed the optically-defined spiral arms, mainly using the positions of blue and red supergiants, and concluded that the plane must be fairly severely warped. They found the inclination to vary from $i=40°$ near the center to 67° at a distance from the center of 30 arcmin. The position angle varied from 50° at the center to 0° in the outer parts. The adopted values for the main disk were $i=44°$ and PA=45°. Considere and Athanassoula (1982), using Fourier analysis of the optical structure, derived values of $i=40°$, PA=35° for the best fit of the bisymmetrical component in the inner 12 arcmin.

Subsequently Maucherat et al. (1984) re-examined the problem using isophotometry, HII region distributions, spiral arm patterns and HI data, concluding that there is no significant warp of the plane and that the "classical" value of the inclination, $i=55°$, is correct (Fig. 4.3).

The reasons for the disagreements appear to be related to the methods used to derive them and the range of radii considered. The spiral structure considerably influences the measured ellipticities of isophotes and their influence is difficult to separate from geometrical effects. On the other hand, if the spiral arms are blotted out from the analyses, as some studies have done, or if only the outer distributions of light are used, the results may be influenced by any bulge or halo component of the oldest stars. It is probably safe to use the classical values ($i=40°$, PA=23°), keeping in mind the possible distortion caused by the plane's warp.

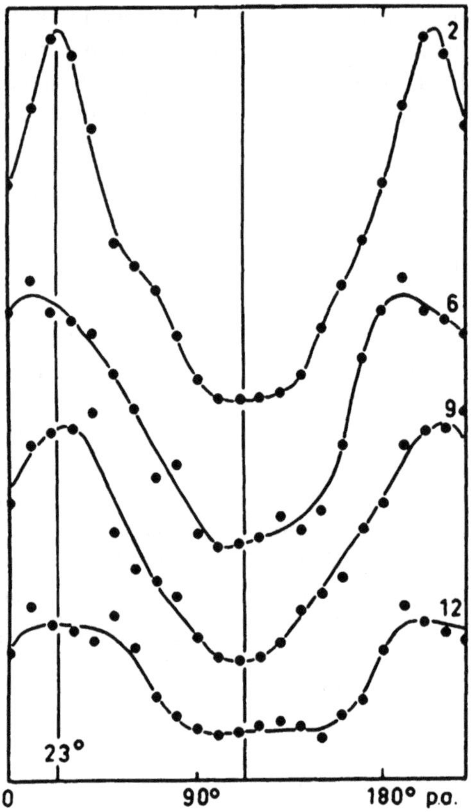

Fig. 4.2 The position angle of the plane of M33 was determined to be 23° from these azimuthal profiles (de Vaucouleurs 1959) (Copyright American Astronomical Society)

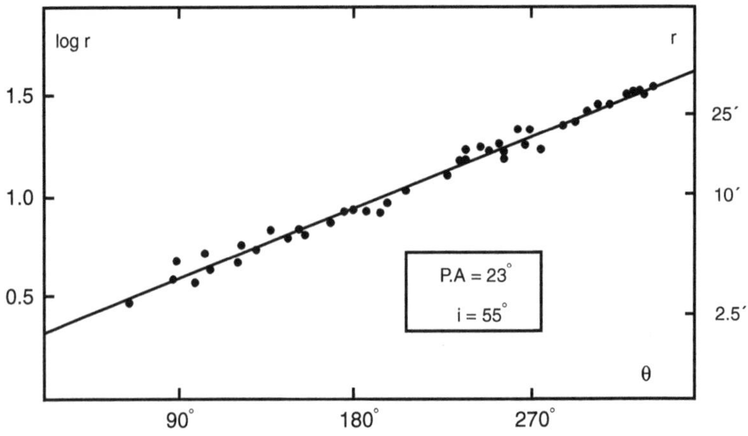

Fig. 4.3 Maucherat et al.'s (1984) plot of the distribution with radius, r, of associations and HI peaks (*points*) compared to a logarithmic spiral (*line*). Used by permission, copyright ESO

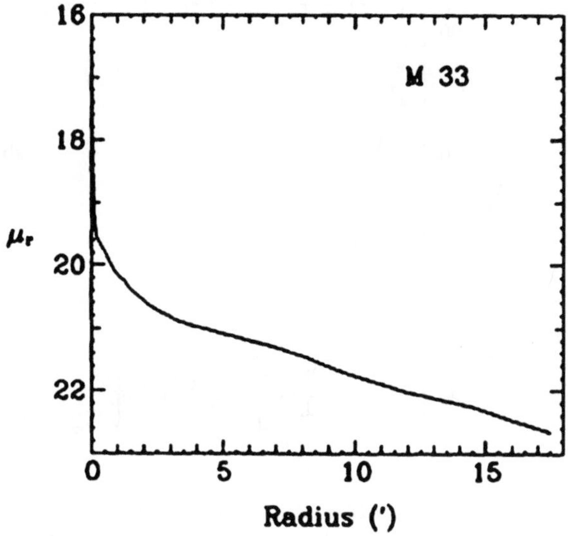

Fig. 4.4 The major axis luminosity profile of M33 in B (Kent 1987). Used by permission, copyright AAS

4.5 Optical Luminosity and Color Profiles

At optical wavelengths the most conspicuous feature of M33 is an exponential stellar disk. Following early photographic studies by Patterson, Shapley, Holmberg and others (loc. cite), the first photoelectric profile was that of de Vaucouleurs (1959), who interpreted his data as showing a disk with an exponential profile with a scale length in B of 1.9 kpc. Later, Kent (1987) derived a scale length of 2.3 kpc (Fig. 4.4).

The diameter of the disk in the optical is a matter of debate. Kent's photometry traced it to 4.3 kpc, but searches for more distant disk stars suggest that it extends considerably farther.

4.6 The Inner Disk and Bulge

Is there a bulge in the inner region of M33? This question has a long history of conflicting claims. First reported by Patterson (1940), who made one of the early photometric studies of M33, it was subsequently unnoticed in the scans made by de Vaucouleurs (1959). The blue and near-infrared images published by Walker showed a smooth population of red stars in the central region and Boulesteix (1979) detected a faint bulge component making up only a hundredth the luminosity of the disk component.

The J and F CCD photometry of Kent (1987) showed a slight decrease in the scale length of the central 3 arcmin, which might indicate a small bulge, but Wilson (1990) suggested that a large stellar association near the nucleus might distort the central disk profile. IRAS infrared data at 12 μm was used by Deul (1989) and Bothun (1992)

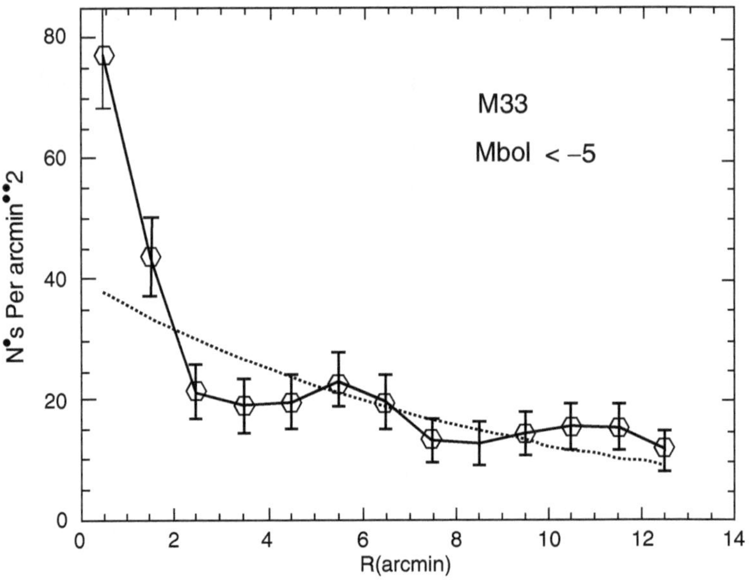

Fig. 4.5 H-band distribution of K and M giants in the M33 bulge (Minniti et al. 1993). The *dashed line* is the disk profile from Kent (1987). Used by permission, copyright AAS

to derive luminosity profiles of the inner region and derived a disk scale length of 1.2 kpc. Bothun's upper limit to the B-band luminosity of the possible bulge was $M_B = -14.15$, but the shape of the profile was not fit well by a $r^{1/4}$ law.

An infrared image of the possible bulge was obtained by Minniti et al. (1993), who obtained an H-band mosaic of M33 which showed a bulge of red stars, identified as K and M giants, located in the inner 2 arcmin of the disk (Fig. 4.5). However, Minniti et al. concluded that this bulge differs from that of the Milky Way Galaxy in that the brightest stars are more luminous, presumably because they are a younger population, less than 10^9 years old.

The controversy did not end with these results. Regan and Vogel (1994) found evidence for a central component with an $r^{1/4}$ profile, but McLean and Liu (1996) measured the IR luminosity profile from 45 arcsec outwards and saw no evidence for any separate component, interpreting observed differences as caused by image crowding and confusion.

Among the spectacular early results of infrared imaging using the Gemini North 8-m telescope was the analysis of the central region of M33 carried out by Stephens and Frogel (2002). They used adaptive optics instrumentation and achieved excellent effective seeing; for instance, the K' band images averaged 0.13 arcsec FWHM. From their observations in the J, H and K bands, they derived color-magnitude diagrams that clearly showed a mix of populations. Analysis showed that the bulge has two components of different age, one formed 0.5 Gyr ago and one 2 Gyr ago. The mean metallicity was found to be moderately high, with [Fe/H] = −0.26 +/− 0.27. Combining the Gemini data with HST imaging at longer wavelengths indicated that the inner stars of M33 represent three populations, a core, a bulge and a disk (Fig. 4.6).

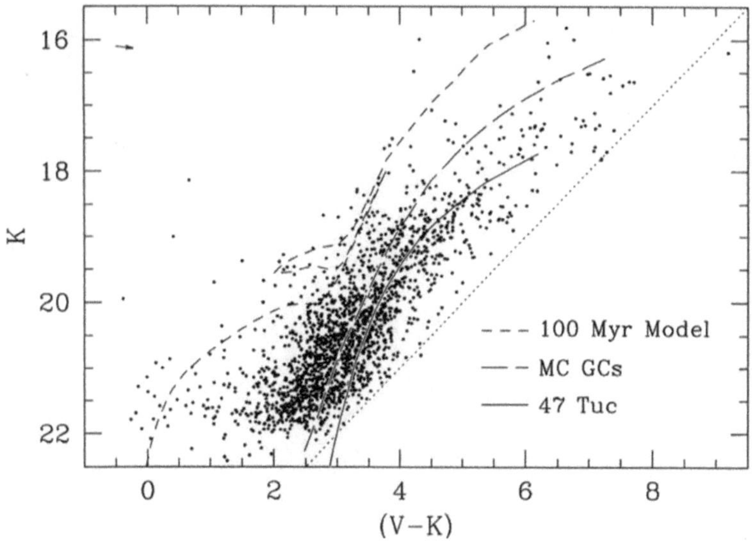

Fig. 4.6 The infrared color-magnitude diagram of the bulge of M33 compared to a model and to globular clusters (Stephens and Frogel 2002). Used by permission, copyright AAS

4.7 The Outer Disk and Halo

The stellar outer disk is spatially separated from the HI disk by at least 30° and the relationship between these is discussed by many papers (e.g. Corbelli and Schneider 1997). Block et al. (2007), using infrared images, found the outer disk to show an enhanced number of carbon stars in an arc at a projected distance of 5 kpc from the center. They hypothesize that it indicates a period of recent star formation, resulting from gas accretion from the HI warp "reservoir". Recent effort has measured the stellar content of the outer disk, as discussed in Chaps. 8 and 9.

The structure of the halo of M33 has only recently been explored in detail. Whether a halo existed was unknown for some time, especially because very few luminous globular clusters like those of M31 were in evidence. A sampling of a distant field, 5 kpc in projected distance from the nucleus, provided the first CMD of the halo (Mould and Kristian 1986). The CMD resembled that of a low metallicity globular cluster. Additional evidence for a traditional Population II halo came from the discovery of RR Lyrae stars (Pritchet 1988) and the halo's kinematics was revealed by the radial velocities of globular clusters (Schommer et al. 1991). Chandar et al. (2002) analyzed the cluster population further and separated disk clusters from halo clusters on the basis of their kinematics (for more about the cluster analysis, see Chaps. 6 and 12).

McConnachie et al. (2006) reported on a survey of the outer parts of M33 using the Isaac Newton Telescope wide-field camera. They detected candidate halo stars far out along the major axis and observed a selection of them spectroscopically with

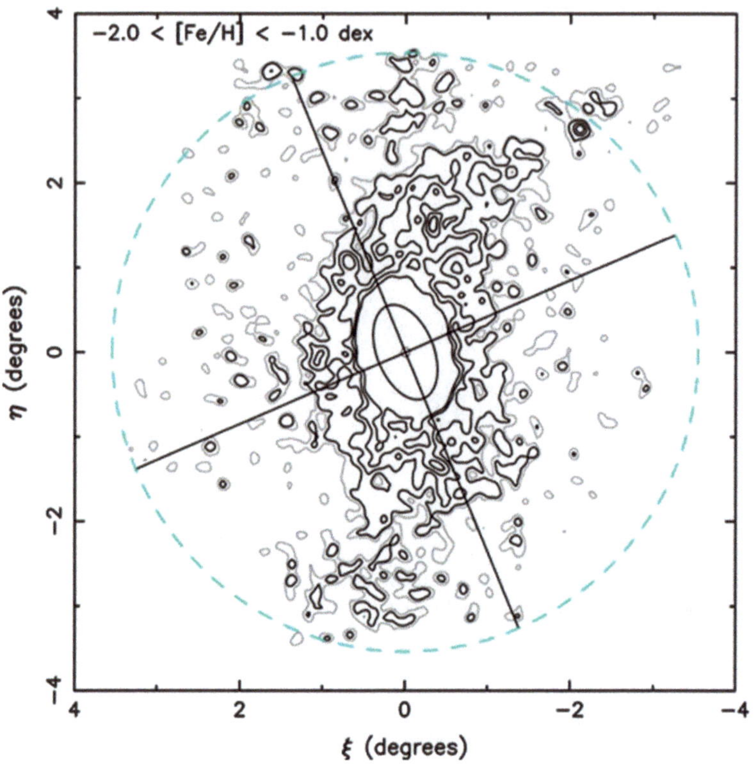

Fig. 4.7 The extreme outer parts of M33. The contours show the distribution of metal-poor giants out to ~40 kpc from the center of M33. The contours are for 1, 2, 5, 8, and 12 σ above the background (From McConnachie et al. 2010). Used by permission, copyright AAS

the Keck II telescope. The radial velocity distributions were found to indicate three different kinematic systems: a halo population with a dispersion of ~ 50 km/s, a disk population with a dispersion of ~ 16 km/s and a third component, identified as a possible tidal feature.

Further exploration of the outskirts of M33 by the same team (McConnachie et al. 2010) extended the range to an area of 40 square degrees, which was imaged in g and i filters with the Canada-France-Hawaii Telescope. Stellar photometry demonstrated the presence of red giant stars out to projected distances from the center of M33 of approximately 40 kpc. A roughly spiral shape was found for the outer stellar distribution, which showed up most clearly for the low metallicity giants (Fig. 4.7). The total mass of this extension was estimated from its luminosity to be about the same as the more massive dwarf ellipticals in the Local Group. The structure was interpreted as having the characteristics of a tidal feature produced by M31's gravitational interaction with M33's outer parts.

4.8 The Spiral Structure

The spiral arm structure of M33 was one of the first cases examined by early studies of spiral galaxies (Danver 1942). Subjective fitting of apparent spiral arms to mathematical spiral shapes indicated that a reasonable fit for M33 was a logarithmic spiral. However, M33 does not have a perfect global spiral structure, but is, at least partly, an example of a "flocculent" spiral (Elmegreen et al. 2003). Only sections of arms are clearly recognizable in the pattern of HII regions, OB associations, uv-bright stars or other indicators of on-going star formation.

Humphreys and Sandage (1980) mapped the OB associations in M33 and showed that they outlined the spiral structure in several areas, notably in a bright southern arm close-in and a wider more open distant arm in the northeast (Fig. 4.8). More recently Wilson (1991) and Regan and Wilson (1993) have attempted less subjective mapping of OB associations to plot the arm structure. The latter reference analyzes the OB associations in the bright "southern arm", finding two different sets of associations, distinguished by an age difference of 2 Myr and a spatial separation of ~1 kpc.

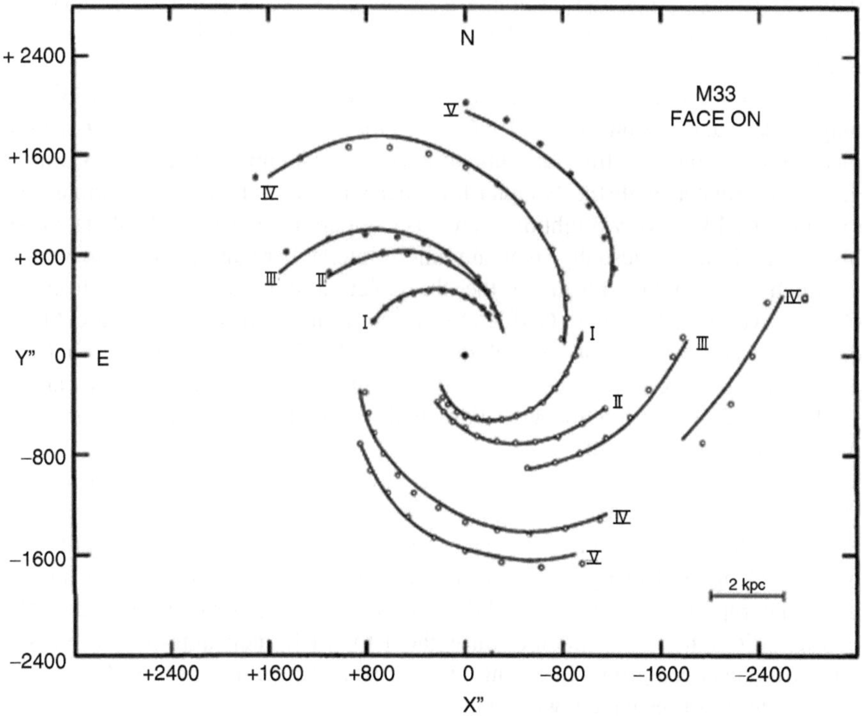

Fig. 4.8 The spiral structure as deduced from the OB associations identified by Humphreys and Sandage (1980). Used by permission, copyright AAS

To eke out the important physical characteristics of M33's optical structure, Elmegreen et al. (2003) analyzed azimuthal scans of an image obtained with the wide-field camera of the CFHT, obtained by Cuillandre, Lequeuex and Loinard (1999). They then obtained Fourier transform power spectra and fitted them to fractal models of stars and continuum. The models provided a best fit of –0.7 +/– 0.7, shallower than expected for a turbulent origin for the structure. However, the authors suggest that foreground stars may result in a spurious shallowing of the slope.

4.9 The Ultraviolet Structure

Global images of M33 in the ultraviolet were obtained with the Ultraviolet Imaging Telescope flown aboard the Space Shuttle Columbia in 1990 (Landsman et al. 1992). Filters at two wavelengths, 152 and 249 nm, allowed both surface photometry and color variations to be traced (Fig. 4.9).

The profiles showed a roughly exponential decrease with radius, with scale lengths of 7.0' at 152 nm and 5.6' at 249 nm. The color becomes bluer with radius, with a range of ~2.5 magnitudes. The arms were found to be ~0.1 magnitudes bluer than the average color and the interarm regions redder by a small and uncertain amount. These results are reasonable for the case of an enhanced star-formation rate in the arms.

The UIT measured ultraviolet fluxes from 16 bright HII regions in M33; the magnitudes ranged from 9.76 for NGC 604 to ~13.5 for the faintest HII regions measured (for the 152 filter and uncorrected for reddening). The uv flux was in general correlated with the H-alpha flux, but with some notable exceptions. For example, IC 133, a very bright uv source, is comparatively weak at H-alpha, possibly the result of an unusually small amount of interstellar material in the nebula.

Much more detailed results on the ultraviolet structure of M33 came from the Galaxy Evolution Explorer (GALEX) spacecraft launched in 2003. The GALEX data represent a major break-through in the uv study of star forming regions in M33, and in spiral galaxies in general, because its relatively high resolution (5 arcsec FWHM) made it possible to study individual young clusters and to map them and their environment in detail.

Thilker et al. (2005). analyzed both far-uv (FUV; 135–175 nm) and near-uv (NUV:175–275 nm) GALEX imaging data. The general structure of the galaxy in the uv is illustrated in Fig. 4.10, which provides luminosity and color profiles for both wavelengths. This figure is based on median surface brightnesses, which minimizes the importance of individual bright sources as compared to mean values. The two uv colors are nearly parallel throughout the disk, indicating that recent star formation has been global in extent, unlike, for instance, the case of M31, where star formation varies widely with radius.

There is only a slight color variation in M33, again in contrast to M31's case where the uv color ranges by almost a magnitude from red in the center to bluer in

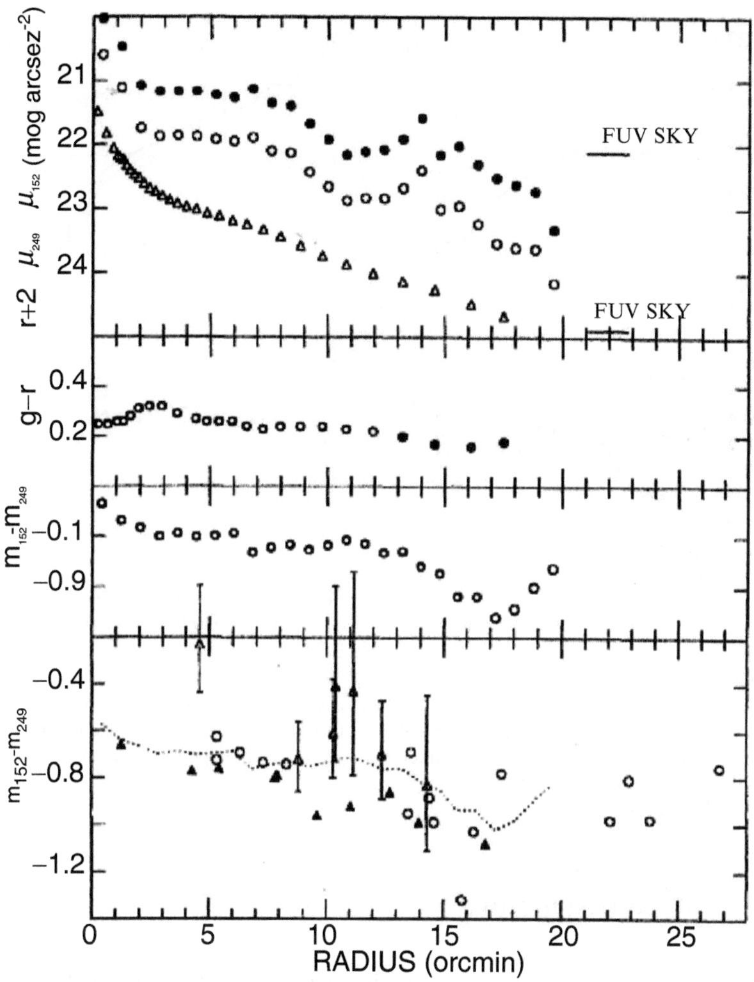

Fig. 4.9 Luminosity profiles for M33 for two wavelengths, FUV, centered at 152 nm (*filled circles*) and NUV, centered at 249 nm (*open circles*). Triangles represent the r-band profile measured by Kent (1987). These profiles were obtained by the space-borne Ultraviolet Imaging Telescope in 1990 (From Landsman et al. 1992). Used by permission, copyright AAS

the 10 kpc ring. HII regions in M33 are found to occur in and near the uv-bright sources, as expected for star-forming regions. There are some instances of bright HII regions where the correlation fails and, though the H-alpha emission's radial profile generally follows the uv emission's, there are regions where they depart from one another, probably because of different extinction and different recent star formation histories.

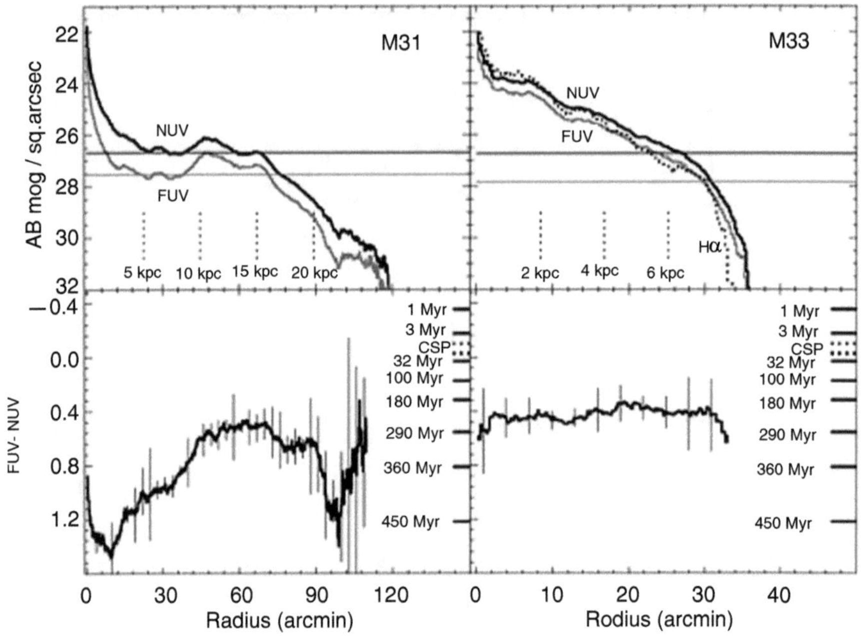

Fig. 4.10 Profiles showing the median surface brightness at both uv wavelengths in M33 and M31. The M33 H-alpha profile is shown as a dotted line. The lower panels show the color curves, where on the right are model ages for single-age populations (*the dotted markers* indicate colors for continuous star formation models) (From Thilker et al. 2005). Used by permission, copyright AAS

4.10 Infrared Structure

Near infrared. The near infrared emission generally traces stars and the effects of HII regions and high-luminosity stars, while the far infrared traces dust at different temperatures. First we discuss the near-infrared structure of M33.

High resolution maps of the infrared emission from M33 at near-infrared wavelengths came from observations with the Kitt Peak Observatory 1.2-m Telescope at the J, H, K and L bands (Regan and Vogel 1994). For the inner 15 arcmin, this study derived profiles that showed an inner spheroidal component and an exponential decrease beyond 3 arcmin (see Fig. 4.11). Scale lengths at J, H, and K averaged 6 arcmin. This is shorter than for the optical scale lengths partly because of a radial variation of extinction.

Far infrared. The first detailed maps in the far infrared were obtained with the IRAS satellite (Rice et al. 1990) at four wavelengths, 12, 25, 60 and 100 μm. These provided resolution at M33 ranging from 180 to 1200 pc, sufficient to study the large-scale structure and a few bright sources. Figure 4.12 provides surface brightness contour maps at these four wavelengths.

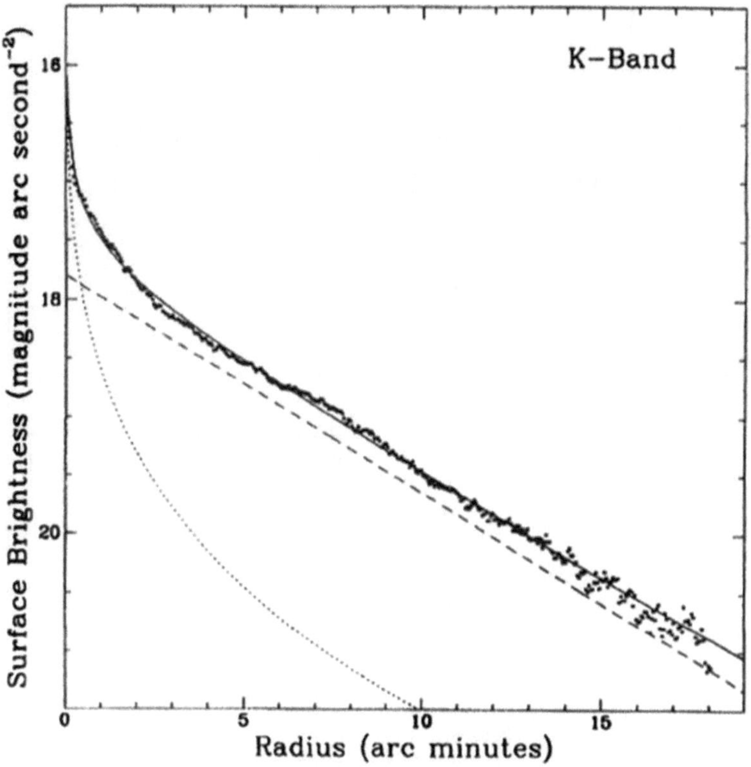

Fig. 4.11 The profile of the H-band emission from M33 (*small dots*). The *dashed line* is the deconvolved exponential component, the dotted line the spheroidal component and the solid line the composite, fit to the observations (From Regan and Vogel 1994). Used by permission, copyright AAS

The structure of the disk in the infrared was derived from the IRAS map using elliptical annuli and solving for the median brightness for each annulus. Figure 4.13 reproduces the IRAS profiles for two wavelengths, 12 and 100 μm. The slopes are different, the shorter wavelength profile being steeper. Scale lengths were 5.3 +/− 0.2 and 7.0 +/− 0.2 arcmin, respectively. By comparison, the scale length at B (also shown in Fig. 4.13) is 8.5 arcmin (de Vaucouleurs 1959). This difference is interpreted as being primarily the result of the fact that the blue light traces stars, with the luminosity modulated by extinction, and the far-infrared light comes from interstellar dust. All profiles at infrared and visual wavelengths decrease exponentially, with bright HII regions and spiral arms producing local peaks. The differences in the infrared profiles can be explained by the different ratios of cool and warm dust, as described in Chap. 6.

The Spitzer infrared space telescope provided even better structural detail for M33 (Gehrz, R. D. et al. 2005).

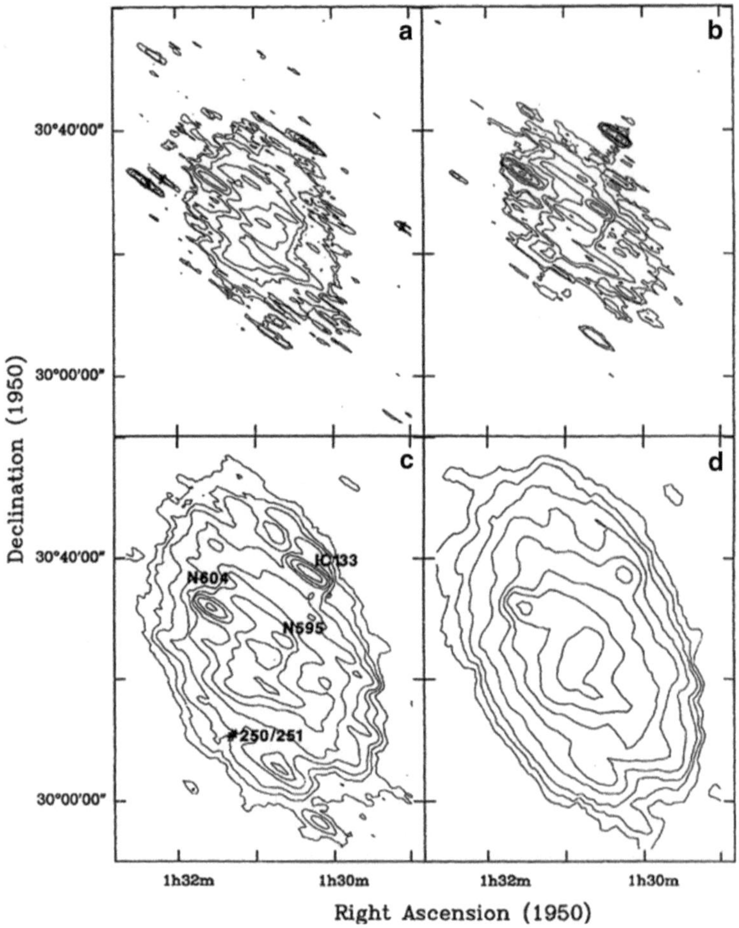

Fig. 4.12 Surface brightness maps of M33 obtained by IRAS. Wavelengths are (**a**) 12, (**b**) 25, (**c**) 60 and (**d**) 100 μm and contours are stepped logarithmically by 2 (From Rice et al. 1990). Used by permission, copyright AAS

4.11 HI Structure

The HI structure of M33 is quite different in several ways from the optical and infrared structure. Most conspicuously, it is greater in extent, it has a shallower profile and it shows an outer warp with respect to the plane of the stellar disk.

Early neutral hydrogen observations of M33 were published by Dieter (1958), who showed that HI is present throughout and beyond the optical disk. As resolution grew better the details of the structure improved, of course. Among many studies, that of Deul and van der Hulst (1987) was a good example of a study of the main body and that of Huchtmeier (1978) pioneered the coverage of the outer areas.

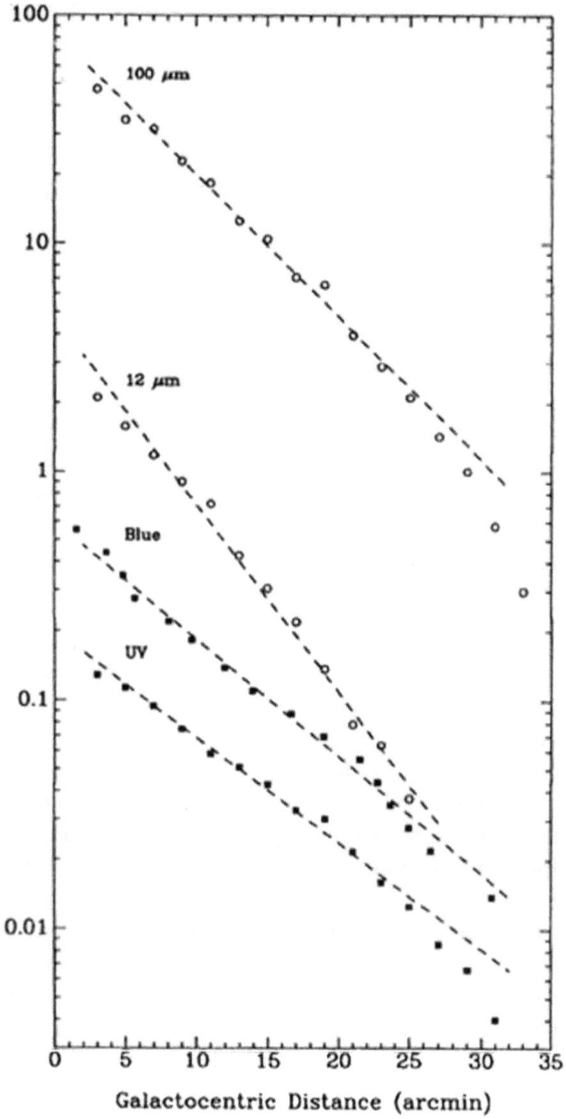

Fig. 4.13 Far-infrared profiles for IRAS wavelength bands, compared to blue and ultraviolet profiles (Rice et al. 1990). Used by permission, copyright AAS

Instead of an exponential decrease in luminosity outwards, the HI profile for M33 is nearly constant for the main body of the disk, with large variations with characteristic scales of ~0.5 kpc (Fig. 4.14). Beyond the optical "edge" at 6 kpc the HI projected density falls off rapidly.

The basic structure in the main disk shows a pattern of arcs and ridges similar to that seen optically as spiral arm segments and that leads to the galaxy's being described as a flocculent-type spiral (Fig. 4.15).

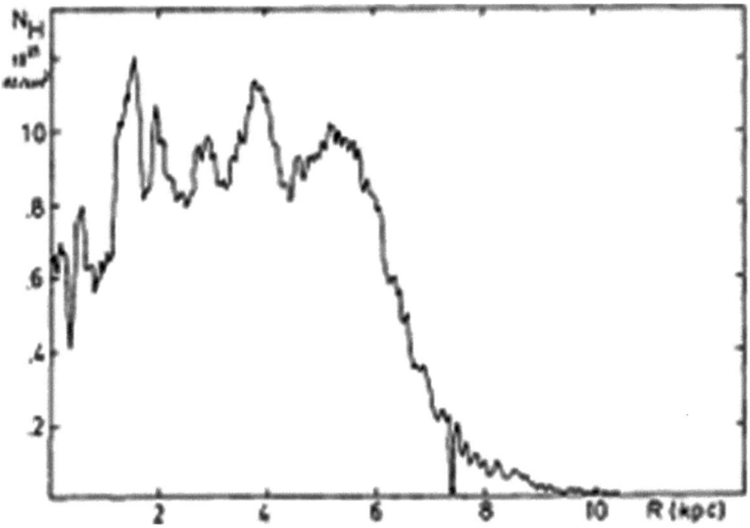

Fig. 4.14 An HI profile of M33 (From Deul and van der Hulst 1987). Used by permission, copyright ESO

In addition to the spiral arm segments, also superimposed on the HI disk are a large number of HI "holes", similar to those found in other late type galaxies. These holes are essentially devoid of emission. They range in size from 40 pc to 1 kpc and have swept-up masses of from 10^3 to 10^7 solar masses. Derived expansion velocities indicate ages less than a few million years and the energies needed to form them range up to 10^{54} ergs. The smaller holes are correlated in position with HII regions and OB associations, while the largest are located in interarm regions and have HII star-forming regions along their edges. The most likely explanation of the smaller holes is that they were formed by supernovae. If so, the implied supernova rate is 1/300 years, a reasonable value compared to values determined from supernova statistics.

Bosma (1981) used aperture synthesis HI data to detect and measure a tilted structure for the outer disk. Using more recent data obtained from the Aricibo 305 m radio telescope, Corbelli and Schneider (1997) produced a similar model of the warp of the HI disk. They found a good fit to the data for a three-ring model that has an outer plane that is oriented ~30 ° from the plane of the inner disk. The best-fit model has about 25% of the HI mass contained in the outer disk, which extends to twice the radius of the optical disk. Their analysis also showed a suggestion of a tidal distortion of the outer disk by M31.

4.12 The Radio Continuum

Radio radiation from M33 was first detected by Terzian and Pankonin (1972). Subsequent study at increasing resolution and sensitivity showed that the disk and sections of the spiral arms emitted diffuse emission and that there were large

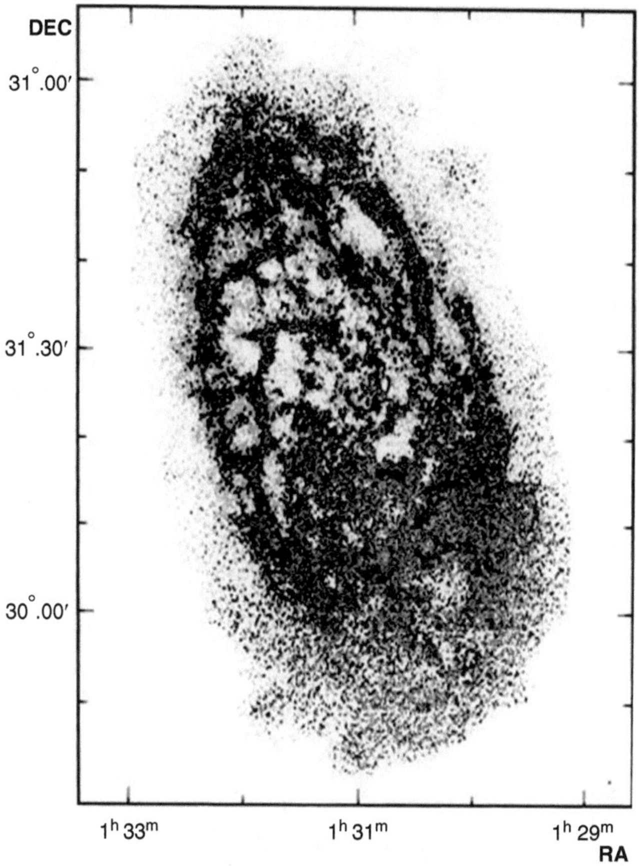

Fig. 4.15 A 21-cm map of M33 (From Deul and van der Hulst 1987). Used by permission, copyright ESO

numbers of discrete radio emitters, many of which were eventually identified as HII regions and SNRs (e.g. Berkhuijsen 1983).

In a series of landmark papers by Tabatabaei et al. (2007a, b, 2008) the then current status of radio continuum studies was presented in detail. They derived radio maps at 3.6 and 6.2 cm using the 100-m Effelsberg Telescope. Figure 4.16 displays the 3.6 cm map and compares it with a superimposed sketch of the spiral arms as derived from OB associations.

Strong radio emission was detected from the main spiral arms and also from some of the minor spiral segments. There appeared to be no clear concentration toward the center of the galaxy, with a nearly flat radial distribute out to R = 4 kpc. Beyond that distance, which is at the outer edge of the main disk in the optical, the emission diminishes more rapidly with radius, with a mean scale length of 1.3 kpc, compared to values near 3.5 kpc inside 4 kpc (at 3.6 cm) (Table 4.1).

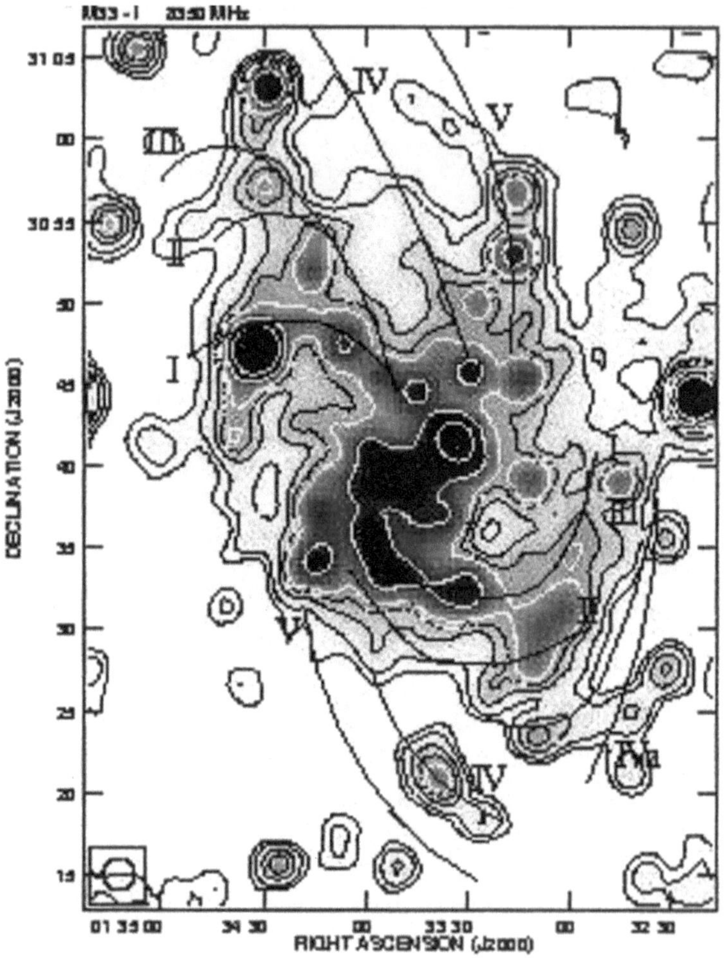

Fig. 4.16 A map of the continuum 3.6 cm radiation from M33. The *curved lines* outline the main optical spiral arms (From Tabatabaei et al. 2007a). Used by permission, copyright ESO

Table 4.1 The radial distribution of continuum emission from M33 (After Tabatabaei et al. 2007b)

Wavelength	Scale length (kpc) for R<4kpc	Scale length (kpc) for R>4 kpc
20 cm	5.4	2.7
6.2 cm	3.7	2.4
3.6 cm	3.5	1.3
24 μm	2.1	1.4
70 μm	2.0	1.8
160 μm	2.5	1.6

The radio spectrum for the emission from the entire galaxy is found to have a mean spectral index of 0.72 +/– 0.04, a value that agrees with those obtained for other Sc galaxies. The spectral index showed a difference for the regions inside vs. outside the R = 4 kpc boundary, with a radial dependence that increased faster in the outer regions. Separating the non-thermal emission from the thermal emission can in principle be a simple task, but in practice it is fraught with complications. The interplay between the different kinds of sources (e.g. hot stars, supernovae, black holes), the nature of the medium (e.g. dust, gas, cosmic rays) and their spatial distribution in the galaxy leads to a complex problem. Early models assumed that the spectral index for nonthermal emission was constant and therefore that the nonthermal emission could be subtracted from the total emission on that basis. However, there are problems with the assumption in detail, as sources of nonthermal emission, e.g. hot stars, supernova remnants and black holes, emit radiation with a spectral index that changes as it propagates out into the galaxy because of energy losses.

An alternative method is to measure the thermal emission by using measures of the optical HII region emission, corrected for reddening, to determine the free-free emission at radio wavelengths. This can then be subtracted from the total radio emission to leave the non-thermal component. An example of the distribution of thermal radiation in M33 is the map shown in Fig. 4.17, which was formed by the use of H-alpha maps, ir maps and radio maps at several wavelengths.

Most of the thermal radiation from M33 comes from luminous HII regions (see Chap. 6), although some also is emitted from the diffuse interstellar medium. The two very bright thermal sources in Fig. 4.16 are NGC 604 (northeast of the center) and NGC 595 (to the west of the center). These two HII regions emit both thermal and nonthermal radio radiation; in both cases the thermal component is about 70% at 3.6 and 20 cm. The overall 3.6 cm radiation from M33 for R < 7.5 kpc is 50% thermal, while for 20 cm radiation it is 18% thermal (Tabatabaei et al. 2008).

The distribution of nonthermal radiation from M33 is more evenly-distributed than the thermal emission. There are local maxima at the positions of luminous HII regions and from some of the spiral arm segments. The HII regions dominate the pattern at 3.6 cm, while they are less important at longer wavelengths. The variation with position in the galaxy of the spectral index of the nonthermal emission suggests that the relativistic electrons lose energy as they disperse from their points if origin. In a related finding, the radio emission is mostly nonthermal at distances from the center of M33 greater than 5 kpc.

The radio emission from M33 has provided information on its general magnetic field. Polarization measurements can determine the intensity and direction of the magnetic field in both the plane of the sky (from linear polarization) and in the line-of-sight (from Faraday rotation). Beck (1979) discovered that the general magnetic field of M33 is bisymmetric, as determined from polarization observations at 11 and 21 cm. More recently, observations at 3.6 and 6.2 cm by Tabatabaei et al. (2008) determined that the general magnetic field consists of two components: an axisymmetric mode within 3 kpc from the nucleus and a superposition of both axisymmetric and bisymmetric modes beyond 3 kpc. The average magnetic field strength of the regular component was found to be 2.5 μG and the average total field strength was 6.4 μG.

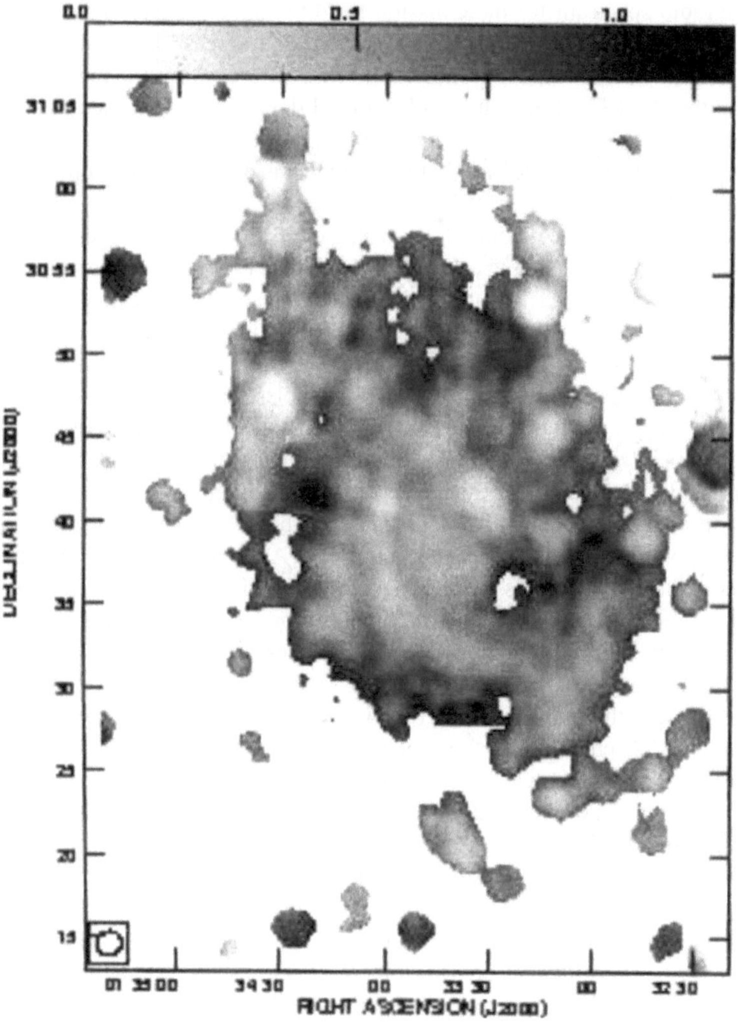

Fig. 4.17 The distribution of thermal radiation emission from M33 at 3.6 cm (After Tabatabaei et al. 2007a). The *grey scale* is indicated at the *top* of the figure, where units are μJy/beam. Used by permission, copyright ESO

4.13 Summing Up

The structure of M33 has been well-explored by measurements and mapping at virtually all available wavelengths. The main stellar structural component is a stellar disk with an exponential radial light profile with a scale length of about 2.0 kpc. There is a possible second component in the form of a bulge, but this is an unresolved question. The exponential disk extends to distances from the center of at least 8 kpc. The slope of the profile is wavelength dependent, being steeper at uv wavelengths than at near infrared. A third component is a halo, which has cold kinematics and which includes stars of a larger variety of metallicities and ages than does our Galaxy's halo. The radio radiation comes from a nearly flat distribution of sources inside a distance of 4 kpc and a steep decline beyond that distance. Recent progress in separating the non-thermal from the thermal radiation has led to a reasonably good understanding of the relative importance of different sources, e.g. HII regions, SNRs, hot stars, collapsed stars and, of course, background sources.

References

Beck, R.: Mitt. Astron. Gesell. **45**, 19 (1979)

Berkhuijsen, E.M.: A&A **127**, 395 (1983)

Block, D.L., et al.: A&A **471**, 467 (2007)

Bosma, A.: AJ **86**, 1825 (1981)

Bothun, G.: AJ **103**, 104 (1992)

Boulesteix, J.: Application de nouvelles methodes digitalisees EN astronomie optique a la physique des regions H II et a la dynamique de la galaxie spirale M33. Univ.Obs, Aix-Marseille (1979)

Chandar, R., Bianchi, L., Ford, H.C., Sarajedini, A.: ApJ **564**, 712 (2002)

Considere, S., Athanassoula, E.: A&A **111**, 28 (1982)

Corbelli, E., Schneider, S.E.: ApJ **479**, 244 (1997)

Cuillandre, J.C., Lequeux, J., Loinard, L.: The stellar content of Local Group galaxies. In: Proceedings of the 192nd Symposium of the International Astronomical Union held in Cape Town, South Africa, 7–11 September 1998. San Francisco, Astronomical Society of the Pacific (ASP) (1999)

Danver, C.G.: Ann. Obs. Lund. **10**, 7 (1942)

de Vaucouleurs, G.: ApJ **130**, 728 (1959)

de Vaucouleurs, G.: ApJS **8**, 31 (1963)

Deul, E.R.: A&A **218**, 78 (1989)

Dieter, N.: AJ **63**, 49 (1958)

Deul, E.R., van der Hulst, J.M.: A&AS **67**, 509 (1987)

Elmegreen, B.G., et al.: ApJ **593**, 333 (2003)

Gehrz, R.D., et al.: BAAS **27**, 451 (2005)

Holmberg, E.: Lund Medd. Astro. Obs. Ser. II. **136**, 1 (1958)

Hubble, E.P.: Realm of the Nebulae. Yale U Press, New Haven (1936)

Huchtmeier, W. K.: Structure and properties of nearby Galaxies. In: Proceedings of the Symposium, Bad Muenstereifel, West Germany, 22–26 August 1977 (A79-13481 03–90), pp. 197. Dordrecht, D. Reidel Publishing Co (1978)

Kent, S.: AJ **94**, 306 (1987)

Landsman, W.B., et al.: ApJ **401**, 83 (1992)

Maucherat., et al.: A&A 133, 341 (1984)

McConnachie, A., et al.: ApJ **647**, L25 (2006)

McConnachie, A., et al.: ApJ **723**, 1038 (2010)

McLean, I.S., Liu, T.: ApJ **456**, 499 (1996)

Minniti, D., Olszewski, E.W., Rieke, M.: ApJ **410**, 79 (1993)

Mould, J., Kristian, J.: ApJ **305**, 591 (1986)

Patterson, F.S.: Harvard Obs. Bull. **914**, 9 (1940)

Pritchet, C. J.: RR Lyrae Stars in Nearby Galaxies. In: Proceedings of the ASP 100th Anniversary Symposium, Victoria, Canada, June 29–July 1 1988 (A90-14129 03–90), pp. 59–67. San Francisco, Astronomical Society of the Pacific

Regan, M.W., Vogel, S.N.: ApJ **434**, 536 (1994)
Regan, M.W., Wilson, C.D.: AJ **105**, 499 (1993)
Rice, W., et al.: ApJ **358**, 418 (1990)
Sandage, A.R.: The Hubble Atlas of Galaxies. Carnegie Inst, Washington (1961)
Sandage, A.R., Humphreys, R.: ApJS **44**, 319 (1980)
Schommer, R.A., et al.: AJ **101**, 873 (1991)
Seyfert, C.K.: ApJ **91**, 528 (1940)
Shapley, H., Ames, A.: Ann. Harvard Obs. **88**, 41 (1932)
Stephens, A.W., Frogel, J.A.: AJ **124**, 202 (2002)
Tabatabaei, F.S., Krause, M., Beck, R.: A&A **472**, 785 (2007a)
Tabatabaei, F.S., et al.: A&A **475**, 133 (2007b)
Tabatabaei, F.S., et al.: A&A **490**, 1005 (2008)
Terzian, Y., Pankonin, V.: ApJ **174**, 293 (1972)
Thilker, D.A., et al.: ApJ **619**, 67 (2005)
van den Bergh, S.: ApJ **131**, 215 (1960)
Wilson, C.D.: AJ **99**, 149 (1990)
Wilson, C.D.: AJ **101**, 1663 (1991)

Chapter 5
The Nucleus

The semi-stellar nucleus of M33 has long been an intriguing if enigmatic feature of that otherwise straightforward galaxy. Sandage and Binggeli (1984) made one of the first measurements of its properties, finding that within an aperture of 6.4 arcsec, the color was $B - V = 0.65$, somewhat bluer than that of a globular cluster. Walker (1964) determined the magnitude of the nucleus within a 10.4 arcsec diaphragm, concluding that $V = 14.54$ for the nucleus, after subtraction of the nearby disk luminosity. This implied an absolute magnitude of $M_V = -10.2$, which made it brighter than any of M33's globular clusters. Walker's values for its color were $B - V = 0.68$, $U - B = 0.05$, which indicated a population with an intermediate average age. Interest in the nucleus increased significantly when it was found that the nucleus coincided in position with M33 X-8, the most luminous persistent X-ray source in the entire Local Group, with an X-ray luminosity of 10^{39} ergs/s (Long et al. 1996). Many studies of the tiny nucleus during the 1990's resulted in data that defined its physical characteristics, though questions of its nature remained.

5.1 Size

Following the pioneer work of Sandage and Binggeli (1984) and Walker (1964), higher-resolution photometry of the nucleus was carried out with both ground-based observatories and the Hubble Space Telescope. Figure 5.1 shows the blue profile that was obtained under superb conditions from the ground by Kormendy and McClure (1993). Even with a seeing-compensating tip-tilt camera achieving a resolution of 0.186 arcsec, the nucleus was unresolved. The implied core radius was less than 0.4 pc and the central density was calculated to be greater than 5×10^5 M_{sun}/pc^3. These data pointed to a remarkable object, a nucleus smaller and denser than most of those in giant spirals.

Higher resolution observations were clearly required and the Hubble Space Telescope soon provided them. Using exposures obtained with the Planetary Camera of the WFPC2, Lauer et al. (1998) and Dubus et al. (1999) achieved resolution at wavelengths ranging from uv (1,491 Å) to the near-infrared (10,190 Å).

P. Hodge, *The Spiral Galaxy M33*, Astrophysics and Space Science Library 379,
DOI 10.1007/978-94-007-2025-1_5, © Springer Science+Business Media B.V. 2012

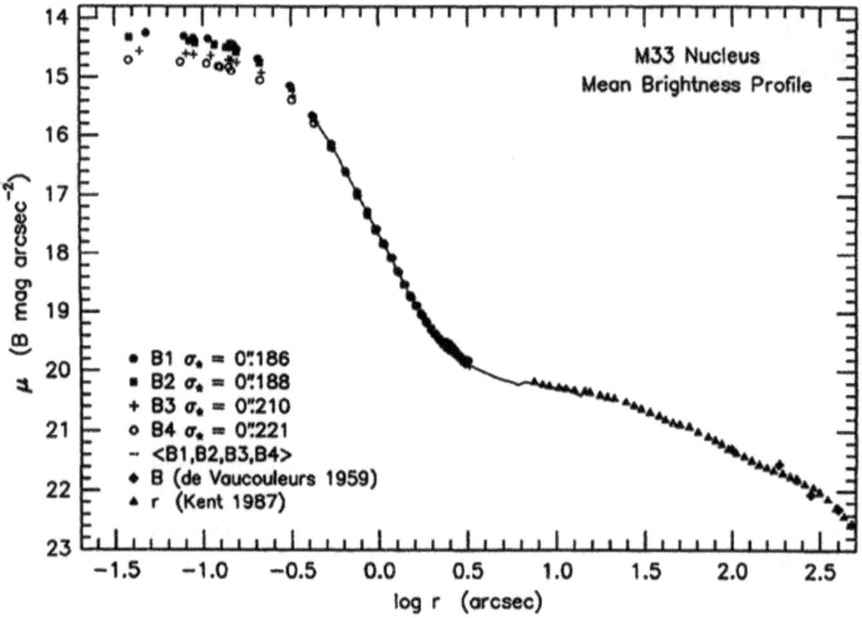

Fig. 5.1 The B-filter profile of the nucleus of M33 obtained at the Canada-France-Hawaii Telescope on Mauna Kea (From Kormendy and McClure 1993). Used by permission, copyright AAS

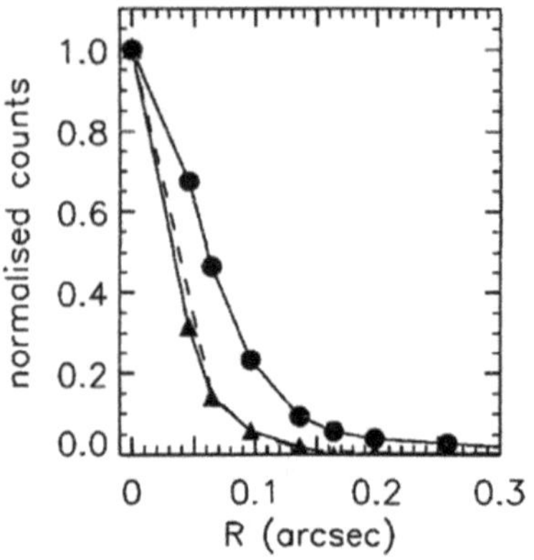

Fig. 5.2 The luminosity profile of the nucleus in the ultraviolet (*dots*), compared to the profile of a nearby star (*triangles*) (From Dubus et al. 1999). Used by permission, copyright AAS

Figure 5.2 shows the profile of the nucleus in the UV. After subtraction of the point-spread-function (Fig. 5.2 shows the scaled profile of a nearby star), the measured FWHM is 0.035 arcsec. This is only 0.14 pc, indicating that M33's nucleus is remarkably compact.

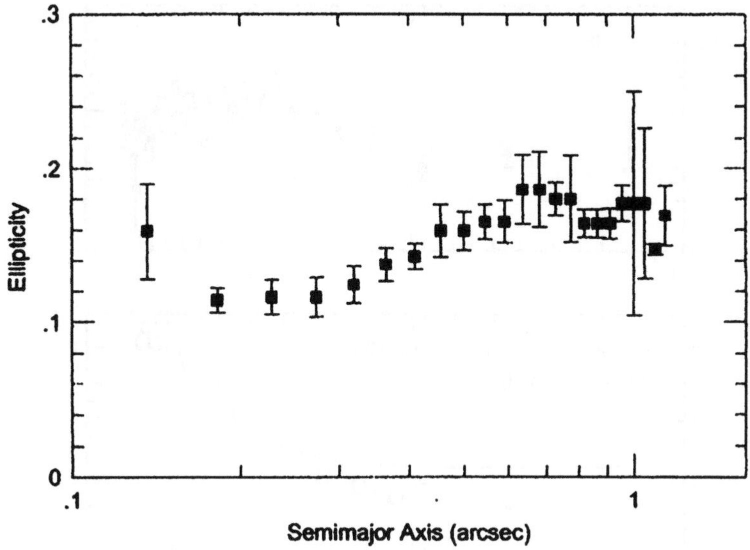

Fig. 5.3 The ellipticity of the nucleus of M33, based on HST images (Lauer et al. 1998). Used by permission, copyright AAS

5.2 Shape

While in luminosity the nucleus is roughly similar to the brightest globular clusters, in other respects there are important differences. One is obvious from the HST images: the nucleus is not spherical. In shape it is flattened, with a mean ellipticity of 0.16 (Fig. 5.3), which is consistent with the fact that most galactic nuclei appear to be flattened systems. Even those late-type galaxies that have larger-scale "nuclear clusters" often show a flattened shape (Seth et al. 2006), a fact that has lead to the hypothesis that nuclear features of late-type galaxies are the result of the concentration of star-forming in-falling gas (or possibly star clusters) funneled in from the outer disk (Kormendy and McClure 1993). Seth et al. (2006) propose a model in which infalling gas first forms stars in a disk, which then expands into an older spheroidal population, all within the nucleus.

5.3 Color

The nucleus is not uniform in color, as was found when it was observed with different sized apertures (van den Bergh 1991). The color was found to be bluer towards the center, ranging from B – R = 0.88 at r = 0.04 arcsec to B – R = 1.25 at r = 3 arcsec (Kormendy and McClure, 1993). A similar result was derived by Lauer et al. (1998), who made measures in V and I (Fig. 5.4).

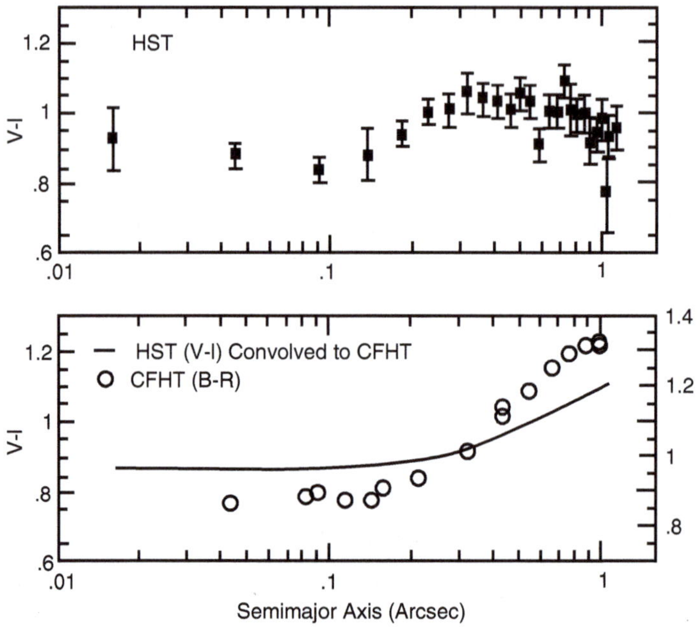

Fig. 5.4 The color profile of the nucleus of M33 based on HST Planetary Camera two images (Lauer et al. 1998). Used by permission, copyright AAS

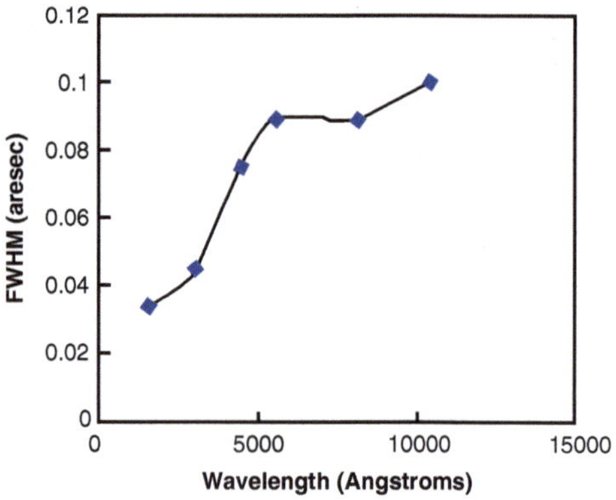

Fig. 5.5 The size of the nucleus (full-width-half-maximum, in arcsec) as a function of the wavelength of the light (From data published by Dubus et al. 1999)

The question of whether the color gradient was the result of abundance differences in the stars or of age differences was settled by the ultraviolet measurements made by Dubus et al. (1999). The very high uv flux indicates that there is a population of very young stars concentrated to the center of the compact nucleus. Approximately 1.5×10^{38} ergs/s of uv light is emitted within 0.14 arcsec of the center. Dubus et al. (1999) show that the size of the nucleus varies over the spectrum from the uv to the near-ir (Fig. 5.5).

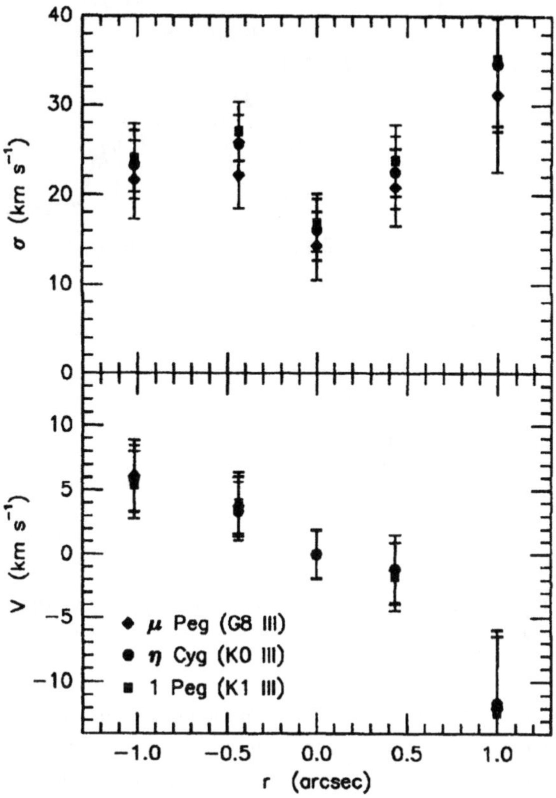

Fig. 5.6 The velocity dispersion (σ) and rotational velocities of the central region of M33, including the nucleus. The symbols indicate the name of the stars used to calibrate the measurements (From Kormendy and McClure 1993). Used by permission, copyright AAS

5.4 Velocity Dispersion

Spectroscopy of the nucleus early-on revealed sharp absorption lines, implying a low velocity dispersion (Gallagher et al. 1982). Kormendy and McClure (1993) measured the velocity dispersion, finding a value of σ=21 +/− 3 km/s (Fig. 5.6). Combined with their measured limits on the size, this implied a small mass-to-light ratio, $<0.4 M_{solar}$. This established the fact that M33's nucleus could not contain a high-mass object like that in the nucleus of M31.

5.5 The Stellar Population

Early photometry and spectroscopy of the nucleus showed that it could not be a globular cluster-like object, as its color was too blue and the spectral energy distribution indicated a population with a mix of ages (Gallagher et al. 1982).

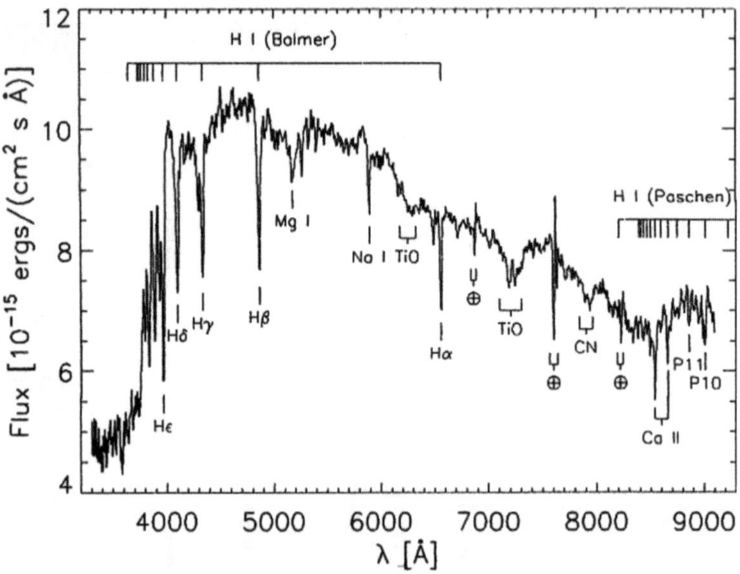

Fig. 5.7 An optical spectrum of the nucleus obtained with the KPNO 2.1-m telescope. Conspicuous lines are identified (From Gordon et al. 1999). Used by permission, copyright AAS

Detailed spectrophotometry of the nucleus using the Lick Observatory 3-m telescope showed that ~50% of the stars were younger than 1 Gyr and that the youngest stars were about 50 million years old (O'Connell 1983, and several others).

More recently, Gordon et al. (1999) obtained optical and ultraviolet spectroscopy and infrared photometry to characterize the population of the nucleus (Fig. 5.7). They concluded that previous studies had been misled by neglecting the effect of dust within the nucleus. Their preferred model was of a population of young stars involved in a dusty environment. The modeling indicated a 70 Myr-old starburst totaling $7 \times 10^5 M_{sun}$ within a shroud of dust with an optical depth of about two. They further concluded that, contrary to previous studies, the dust had a Milky Way-like λ2175 Å bump.

In a program motivated by the desire to identify the source of the strong X-rays, Long et al. (2002) obtained HST STIS spectra of the nucleus and a suspicious-looking blue star just northeast of the nucleus. Their spectra covered the range from 1,150 to 5,700 Å (Fig. 5.8). The blue star turned out to be an ordinary O-type star and thus supposedly not related to the X-Ray source. In agreement with O'Connell's results and contrary to the model of Gordon et al., the data indicated that there had been two starburst events in the nucleus, one about 40 Myr ago and the other near 1 Gyr ago. The more recent involved about 10^4 M_{sun}, while the older starburst involved nearly 8×10^5 M_{sun}.

They found little evidence for a large quantity of dust; furthermore, the 2,175 Å bump was found to be inconspicuous in the spectrum of both the nucleus and the blue nearby star.

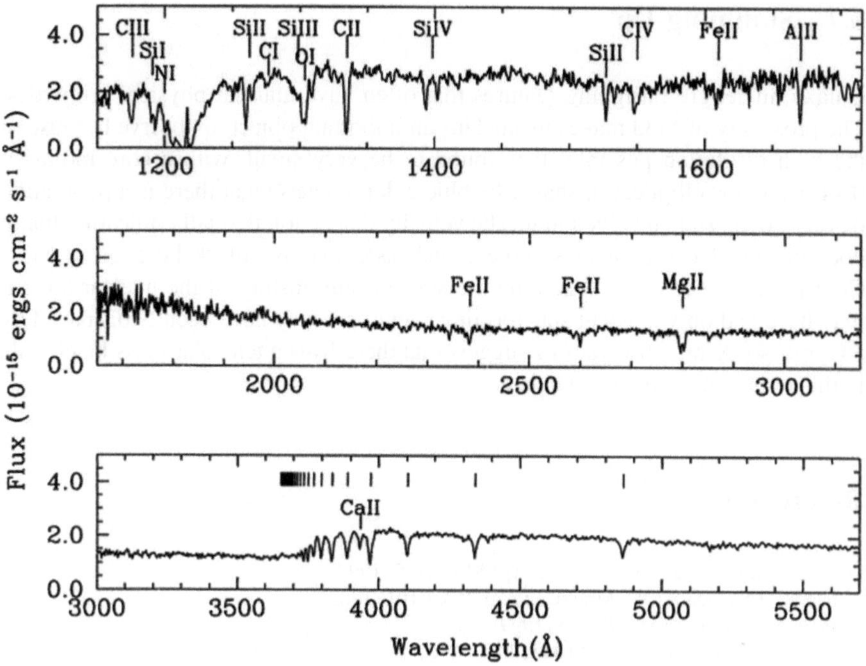

Fig. 5.8 A spectrum of M33's nucleus obtained with the HST STIS spectrograph. Important absorption lines are identified (From Long et al. 2002). Used by permission, copyright AAS

It is not possible to state that we now know the answer to the question of the star-formation history of the nucleus of M33. Was there a single recent starburst in a region enveloped thickly with Milky-Way style dust or were there two bursts, one recent and one a Gyr ago in a relatively dust-free environment? And how does the presence of an unusual X-Ray source influence the interpretation?

5.6 Is There a Black Hole?

All studies of the nucleus since the pioneering work of Kormendy and McClure (1993) agree that there is no indication of a massive black hole in the nucleus. However, the presence of a remarkably bright X-ray source suggests that there is a 10-solar mass black hole binary (see Chap. 11). As far as the stellar population question goes, an evolved star this massive points to the presence of a recent stellar formation event, in agreement with the spectroscopic and photometric evidence.

5.7 Summing-Up

Galaxy nuclei are intriguing features that often have unusual physical properties. The proximity of M33 makes its nucleus an important object to observe because of the high resolution possible. It is found to be very small, with a core radius of 0.14 pc, and is elliptical in shape. Its blue color suggests that there is a population of very young stars involved in it. The velocity dispersion is small, indicating that it does not involve a high mass object such as a massive black hole, as in M31. Contrary evidence exists regarding the star-forming history of the nuclear area; it may have had only a recent star-forming event or it may have been two, spaced by a Gyr or so. X-ray observations suggest that there lives a ten solar mass black hole in this remarkable compact nucleus.

References

Dubus, G., Long, K.S., Charles, P.A.: ApJ **519**, L135 (1999)
Gallagher, J.S., Goad, J.W., Mould, J.: ApJ **263**, L101 (1982)
Gordon, K.D., et al.: ApJ **519**, 165 (1999)
Kormendy, J., McClure, R.D.: AJ **105**, 179 (1993)
Lauer, T.R., et al.: AJ **116**, 2263 (1998)
Long, K.S., et al.: ApJ **466**, 750 (1996)
Long, K.S., Charles, P.A., Dubus, G.: ApJ **569**, L204 (2002)
O'Connell, R.W.: ApJ **267**, 80 (1983)
Sandage, A.R., Binggeli, B.: AJ **89**, 919 (1984)
Seth, A.C., et al.: AJ **132**, 2539 (2006)
van den Bergh, S.: PASP **103**, 609 (1991)
Walker, M.F.: AJ **69**, 744 (1964)

Chapter 6
Clusters and Associations

6.1 Searches and Catalogs

The star clusters of M33, after a slow start, have become a rich and revealing source of information on the galaxy's history and its modes of star-formation. The discovery and analysis of clusters follows, to a large extent, the pattern of discovery of the rich cluster systems of the Magellanic Clouds (Westerlund 1997), but delayed in time by several decades. The first catalog of clusters was published by Hiltner (1960), who found 23 cluster candidates, based on examination of photographic plates taken with the Mt. Wilson 2.5 m telescope. Hiltner followed the discovery with photometry in the UBV system, from which he concluded that M33's clusters were bluer and fainter than those of M31.

It was a surprising 18 years before further progress was made in the survey of M33 clusters, when Melnick and D'Odorico (1978) reported the discovery of 33 more candidates and provided blue magnitudes for them. Four years later Christian and Schommer (1982) listed 250 non-stellar objects in M33, based primarily on a photographic plate taken with the KPNO 4 m telescope. This was followed by photometric and spectroscopic observations for some of them and an analysis of their types and their kinematics (Christian and Schommer 1982; Schommer et al. 1991). Mochejska et al. (1998) carried out a subsequent survey and discussion of M33 clusters, including 35 new candidates.

With the availability of the Hubble Space Telescope, higher definition and deeper exposures allowed better recognition of clusters and fuller possibilities for determining their properties. In a series of papers using HST data, Chandar et al. (1999a, b, c, 2001) reported the discovery of 131 new clusters, for which they provided integrated photometry in various color systems. Bedin et al. (2005) and Sarajedini et al. (2007) used the HST ACS camera to search for additional new clusters. The former reported the detection of 84 candidates on a single ACS image and the latter discovered 12 new clusters and provided photometry for them and for 12 others.

A master database for M33 clusters was constructed by Sarajedini and Mancone (2007), who compiled all of the above sources into a catalog of 451 cluster candidates.

P. Hodge, *The Spiral Galaxy M33*, Astrophysics and Space Science Library 379,
DOI 10.1007/978-94-007-2025-1_6, © Springer Science+Business Media B.V. 2012

Based on ST images and high-resolution ground-based data, they selected 255 objects as high-confidence cluster candidates and analyzed their individual and global properties, all on a transformed common photometric system. An updated version of this catalog was placed on the web, and as of May, 2011 it contained a list of 595 objects.

(http://www.astro.ufl.edu/~ata/cgi-bin/m33_cluster_catalog/index.cgi).

A new ground-based search for cluster candidates was initiated by Zloczewski et al. (2008). They used the CFHT telescope's MegaCam to survey a large area centered on M33 and found 4,780 extended sources, of which 3,554 were candidate star clusters. Several were possible halo objects in the outer parts of the galaxy, including one apparent old globular cluster at a distance of 27 arcmin from the center of M33. In an examination of these candidates on HST images that overlapped, San Roman et al. (2009) found that of the 60 candidates, only 21 objects were resolved into stars. The remaining candidates appeared to be small asterisms, background galaxies and diffuse nebulae. Extrapolating to the complete Zloczewski et al. catalog, San Roman et al. calculated that about 1,400 objects might be true star clusters. This can be compared to the 480 clusters in the Sarajedini-Mancone catalog at that time, of which the authors considered only 264 to be confirmed clusters.

The paper by San Roman et al. (2009) reported the results of a photometric study of M33 clusters located on a series of HST ACS images. A total of 161 clusters was included, 46 of which had been cataloged previous to 2009. Their results included integrated photometry and color-magnitude diagrams. The clusters ranged in absolute magnitude from $M_V \sim -9$ to ~ -4 and ages ranged from $\sim 10^7$ to $\sim 10^9$ years.

6.2 Globular Clusters

It has taken many years to make a separation of the globular clusters in M33 from the non-globular clusters. This is a situation similar to the case of the Magellanic Clouds in the middle of the last century. Much of the problem is due to expectations based on the situation in our own Galaxy, which has (for the most part) a clearer separation between old, metal-rich halo and bulge clusters and those smaller, younger clusters inhabiting the disk. In the Clouds, and, apparently in M33, the clusters seem to inhabit more fully all ranges of magnitude and color, and by inference wide ranges of age. They seem to be distinguishable as globular or non-globular primarily by their kinematics, as a small number of clusters were found by Schommer et al. (1991) to have halo-type orbits. Other candidate globulars were identified as such by their location beyond the apparent disk of the galaxy or by their color-magnitude diagrams. But there are many luminous clusters that have red colors that remain in the area of the CMD that is populated by objects older than ~1 Gyr and, though some may be classical globulars, most are probably what are in the Magellanic Clouds called "young populous clusters", massive clusters of intermediate age, objects comparatively rare in the Milky Way Galaxy.

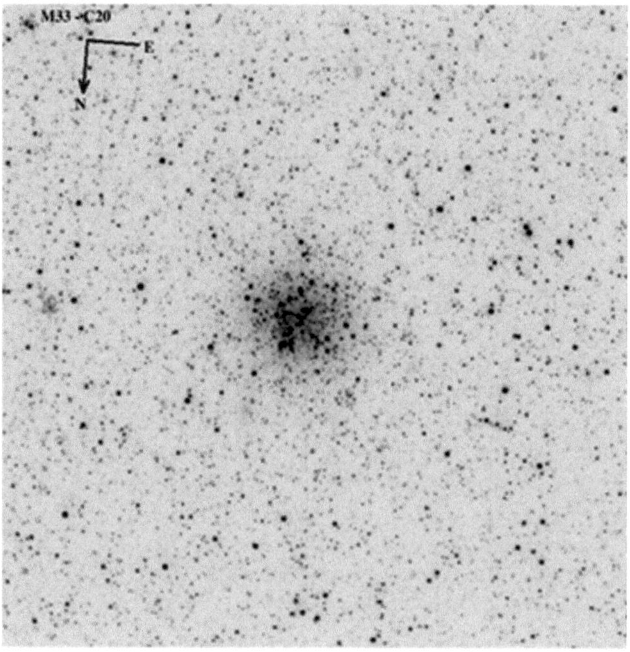

Fig. 6.1 The globular star cluster C20, a metal-poor halo cluster (From Sarajedini et al. 2000). Used by permission, copyright AAS

HST WFPC2 observations of 10 halo clusters in M33 by Sarajedini et al. (1998) showed color-magnitude diagrams that indicated ages younger than globulars in the Milky Way. These clusters all had the structure and high luminosity of typical globular clusters (see Figs. 6.1 and 6.2).

The horizontal branches of eight of them were red, while the metallicities were low, with [Fe/H] ~ – 1.6. Assuming that the color of the horizontal branch correlates with age, the CMDs indicated ages of only a few Gyr. In a subsequent paper (Sarajedini et al. 2000), these authors derived more precise characteristics for the clusters, which were said to have metal abundances of [Fe/H] = –1.64 to – 0.65, based on the slopes of their red giant branches. Ages for those with red HB's averaged ~7 Gyr, while the remaining two were found to be as old as globulars in the Milky Way (Figs. 6.3 and 6.4). From these data, it was clear that M33 had a different cluster-forming history from that of the Milky Way and M31, with the massive clusters being formed steadily after the initial formation of a few clusters shortly after the big bang.

Four of the classical globular clusters, H38, M9, R12 and U49, were studied spectroscopically with the Keck I Telescope and its echelle spectrograph (Larsen et al. 2002). Figure 6.5 shows a section of one of the high-dispersion spectra, compared to a standard template to permit the measurement of the cluster's velocity dispersion.

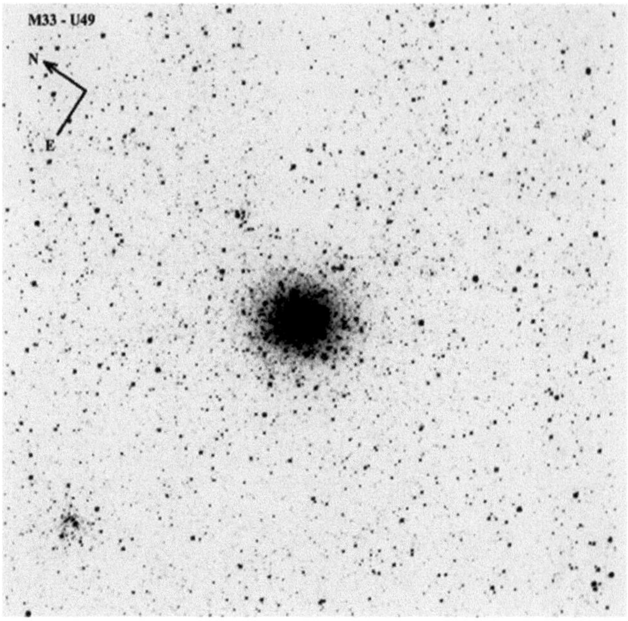

Fig. 6.2 The globular cluster U49, a metal-poor halo cluster (From Sarajedini et al. 2000). Used by permission, copyright AAS

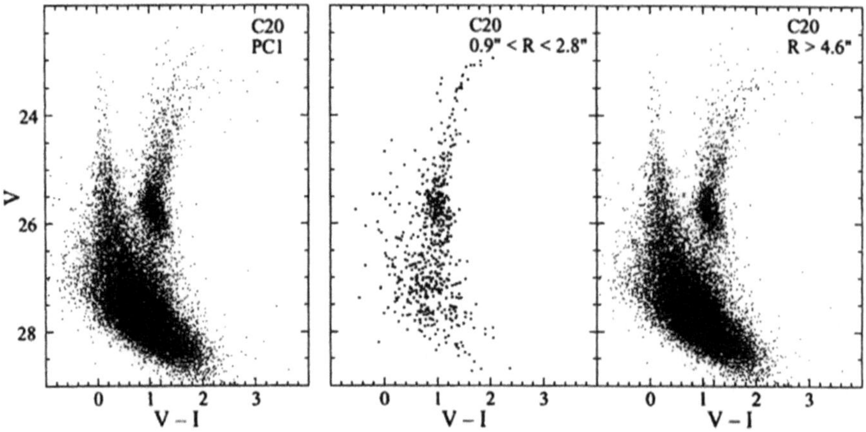

Fig. 6.3 The CMD of the old globular cluster C20. The *central panel* shows the data for the cluster, while the *right panel* shows the stars immediately surrounding the cluster and the left shows all of the stars on the HST Planetary Camera image (From Sarajedini et al. 2000). Used by permission, copyright AAS

The velocity dispersions of the four clusters, combined with model fitting to King-Michie profiles (Fig. 6.6), led to a determination of the clusters' masses. Their mass-to-light ratios had a mean value of 1.53 +/– 0.18 (solar units and V color), which agrees with typical values for Milky Way globulars. Larsen et al. (2002) compared their characteristics with the fundamental plane of Milky Way globular

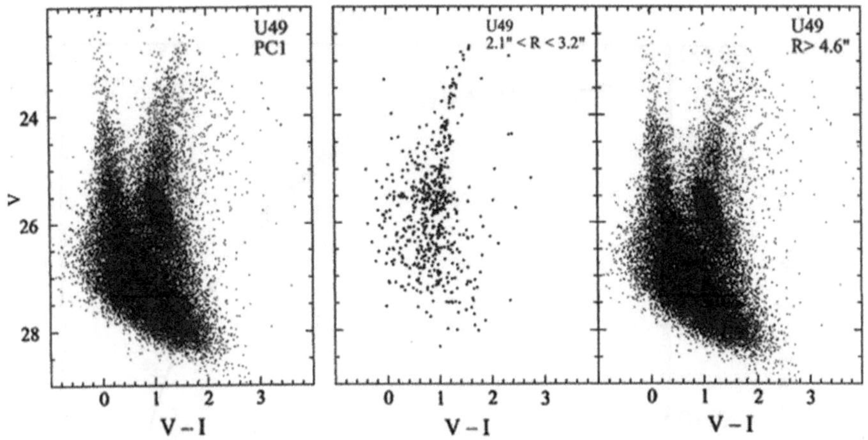

Fig. 6.4 The CMD of the intermediate age cluster U49. The panels are as in Fig. 6.3 (From Sarajedini et al. 2000). Used by permission, copyright AAS

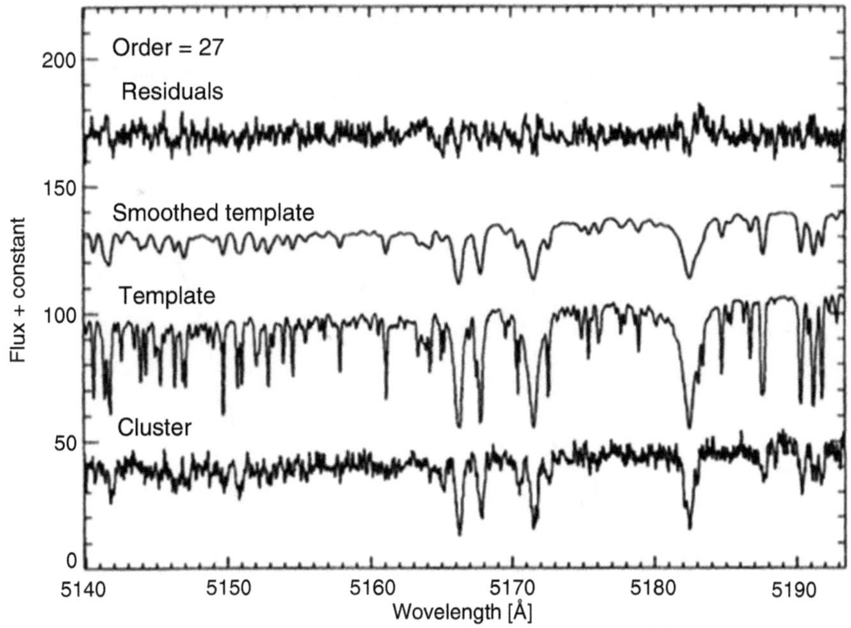

Fig. 6.5 A portion of a high-dispersion spectrum of globular cluster H38, compared to a template. The spectral region includes the calcium triplet (From Larsen et al. 2002). Used by permission, copyright AAS

clusters and with its binding energy-luminosity relation and concluded that the M33 clusters were essentially identical to those of the Milky Way.

Chandar et al. (2006), while spectroscopically observing several apparent classical globulars with the MMT, found that cluster M33-C38 had the same low abundances as the others, but with an indicated age of ~7 Gyr (Fig. 6.7). Its mass was calculated

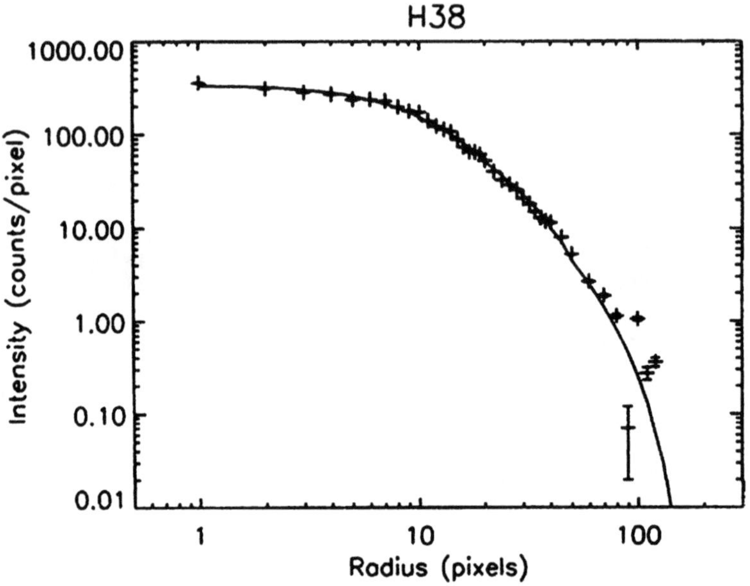

Fig. 6.6 The luminosity profile of the old globular cluster H38 fit to a dynamical King-Michie model (From Larsen et al. 2002). Used by permission, copyright AAS

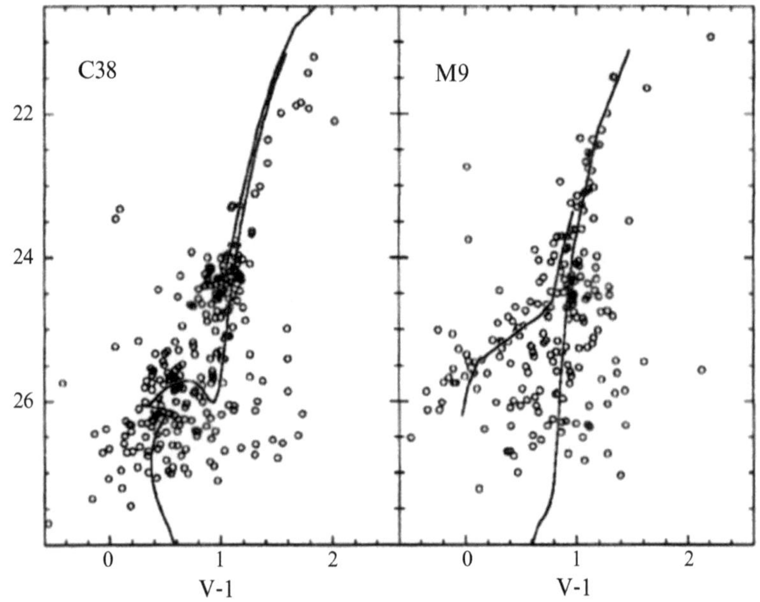

Fig. 6.7 A comparison of the CMD of the old globular M9 with that of the younger cluster C38 (From Chandar et al. 2006). Used by permission, copyright AAS

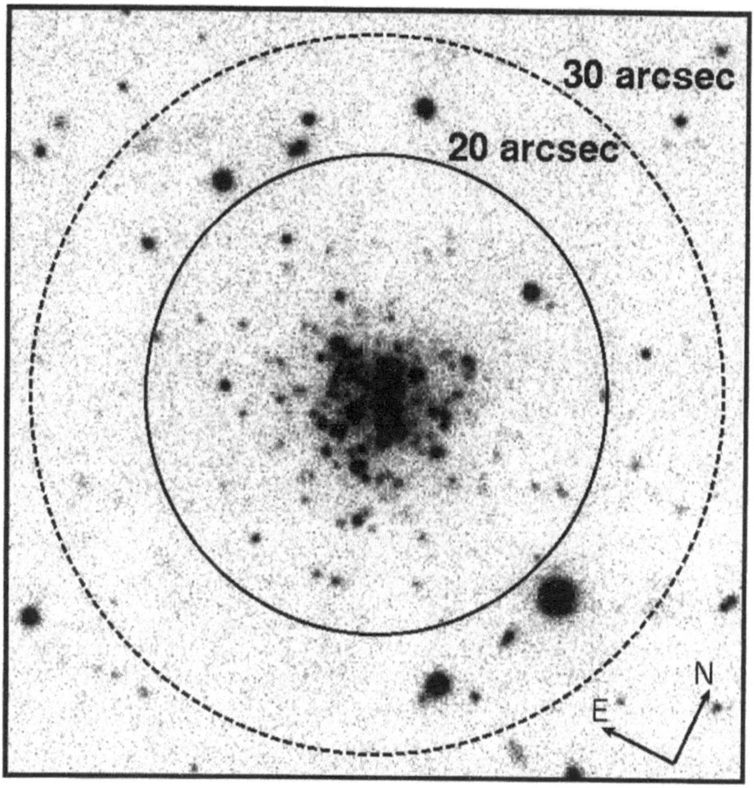

Fig. 6.8 The extended cluster EC1 (From Stonkute et al. 2008). Used by permission, copyright AAS

to be $(0.8 - 1.2) \times 10^5$ solar masses. This case supports previous indications that M33 was able to form massive clusters with low [Fe/H] for several Gyr after the initial burst of cluster formation.

M33 also has at least one "extended" cluster (Fig. 6.8, Stonkute et al. 2008), similar to the large, low-density clusters found in M31. Called M33-EC1, the cluster lies at a projected distance from M33 of 12.5 kpc.

Its color-magnitude diagram indicated an age of about 7 Gyr and a metallicity of – 1.4. Stonkute et al. (2008) suggested that if these objects are the remnants of tidally-stripped dwarf galaxies, as is thought possible, then its existence indicates that M33 may have had a complex merging history (Fig. 6.9).

Six very distant globular star clusters have been discovered in large-scale photometric surveys of the outer regions around M33 (Huxor et al. 2009; Cockcroft 2011). They lie at projected distances greater than 10 kpc, among the vast regions of very thinly-spaced red giants (Chap. 4). Two of them are probably "extended clusters", while the rest have structures like those of normal globulars. All apparently have low to intermediate metals abundances, like the halo clusters found closer to the center.

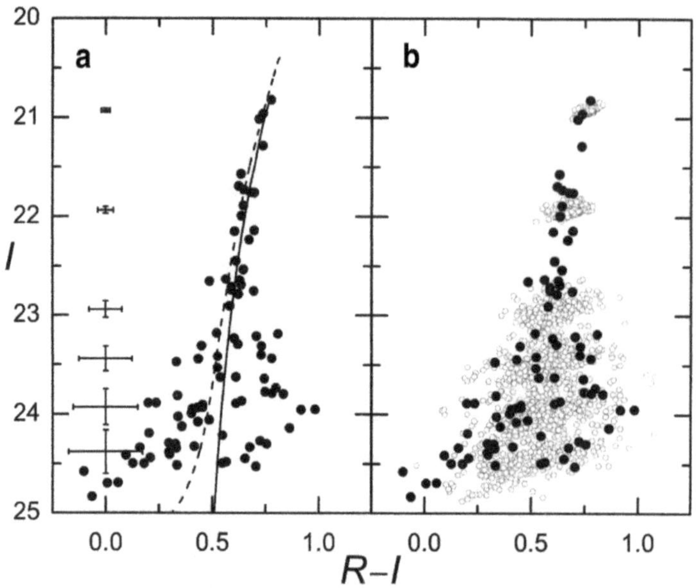

Fig. 6.9 The CMD of the large, low-density cluster EC1, showing a typical globular cluster giant branch. The *left panel* (**a**) shows the cluster CMD superimposed upon an evolutionary model with [Fe/H] = −1.3 and age = 14 Gyr, while the *right panel* (**b**) shows the cluster data against the artificial star data used to determine the error bars in panel a (From Stonkute et al. 2008). Used by permission, copyright AAS

6.3 Young Populous Clusters

The previous section describes the globular-like clusters that inhabit the halo of M33 and are seen against its disk. These are found to range in age from a few to about 12 Gyr. Also found in abundance, but mostly located in the galaxy's disk, are luminous clusters with younger ages. Many of these objects are as luminous as the globulars and are similar to clusters that are common in the Magellanic Clouds, but rarely found in the Milky Way Galaxy. The most luminous are sometimes referred to as "blue globular clusters", but most are not closely related to traditional globulars. They are generally less massive, have higher abundances, are younger and have disk-like kinematics. Prior to 2010, most clusters examined in detail in M33 have been of this type. Their absolute magnitudes range from $M_V = -9$ to −4 and masses range from 5×10^3 to close to 10^5 solar masses. Their colors and inferentially their ages cover the spectrum from old, nearly globular-cluster-ages to blue, very young ages. Figure 6.10 shows a color-magnitude diagram that includes most confirmed clusters of all types, in which most would be considered luminous enough to be called young populous clusters. Some of those brighter than $M_V = -7$ and redder than $V - I = 1.0$ are genuine globular clusters, as determined by their CMDs, but most clusters in Fig. 6.10 are of intermediate age (which has come to mean any age less than ~ 10 Gyr and greater than ~100 Myr).

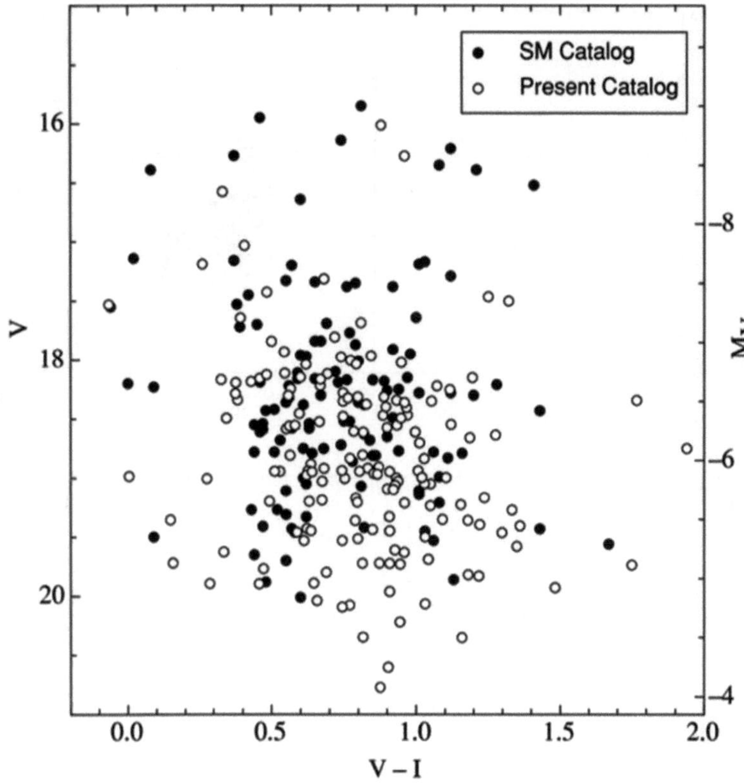

Fig. 6.10 The color-magnitude diagram for confirmed M33 star clusters. *Open circles* are from the Sarajedini and Mancone (2007) catalog and *filled circles* from the San Roman et al. (2009) addition to the catalog. Used by permission, copyright AAS

The cluster CMD plotted in Fig. 6.10 might appear puzzlingly uniform. Considering that the plot is a log-log plot and yet the data, number vs time (inferred ages from colors) are linear, it might be expected to show some, probably extreme, asymmetry. However, this uniformity is just a coincidence. The effects of the logarithmic age/color scale cause the density (in the diagram) to spread out towards the blue and the effects of cluster disintegration and evolutionary fading cause a decrease towards the red from the centroid, compared to the case of a constant formation rate.

Sarajedini and Mancone (2007) pointed out the similarity of the color distribution of M33 clusters to that of the Magellanic Cloud clusters (Fig. 6.11). Both distributions show a Gaussian-like distribution with a maximum population density at $(B-V)_0 = \sim 0.2$ and a range from $(B-V)_0 = -0.5$ to 1.0.

Of course, to determine the ages of clusters it is necessary to acquire more data than their observed colors. Precise ages of clusters in other galaxies are difficult to obtain,

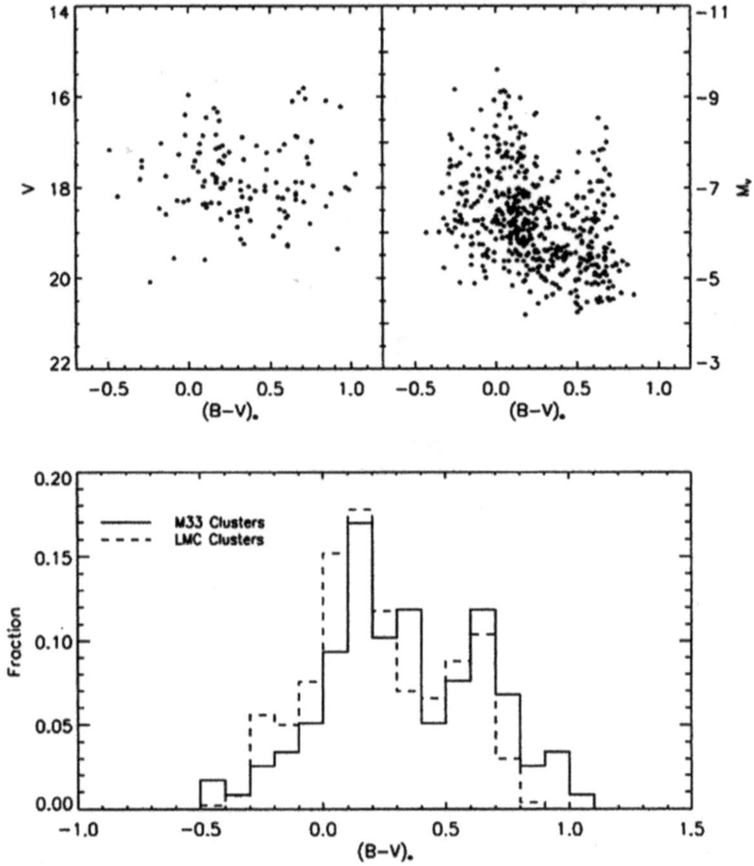

Fig. 6.11 Top: a comparison of the reddening-corrected color-magnitude diagram of M33 clusters (*left panel*) with that for the Large Magellanic Cloud young populous clusters (*right panel*). Bottom: The distribution in color for the two populations of clusters, normalized to unit area (Sarajedini and Mancone 2007). Used by permission, copyright AAS

requiring knowing the reddening, the chemical abundances and the spectral energy distribution. Approximate ages can be found if magnitudes in at least three wavelengths are measured, if the reddening is determined (for example, from the two-color diagram) (Fig. 6.12), if the abundances are assumed or determined from spectra and if particular theoretical stellar evolutionary models are assumed. In this way ages of several samples of M33 young populous clusters were obtained by, for instance, Chandar et al. (1999c) and Sarajedini and Mancone (2007). In all cases it was found that the number of clusters of a given age shows a very steep decrease with age.

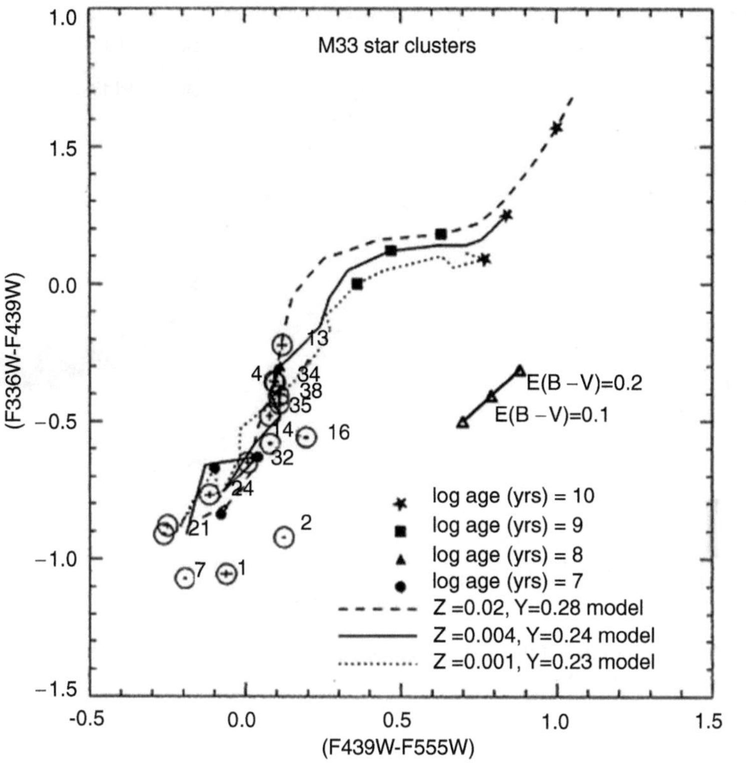

Fig. 6.12 A color-color diagram for young clusters in M33, compared to Padua evolutionary models from Chandar et al. (1999b). Used by permission, copyright AAS

The number of clusters with ages of 5 Myr is four orders of magnitude greater than the number 10 Gyrs old. This kind of relationship is well-known for all galaxies for which the age distribution has been determined and is the result of dynamical evolution (primarily by tidal interactions with GMCs and other massive bodies). Among the many papers on this subject, we note the excellent discussion of the dissolution of clusters in different galaxies, including M33, published by Lamers, Gieles and Portegies Zwart (2005).

The structure of the populous clusters of M33 was investigated by Chandar et al. (1999b), who measured radial profiles for 60 clusters. Examples for clusters R12, LB 4–2 and PM 3–5 are shown in Fig. 6.13. Fifty-one of the clusters fit single mass King (1966) models well, with a mean core radius of 0.7 pc and a range from 0.12 to 1.57 pc. This can be compared to the mean core radius for Milky Way globular clusters, which range from 0.10 to 27 pc.

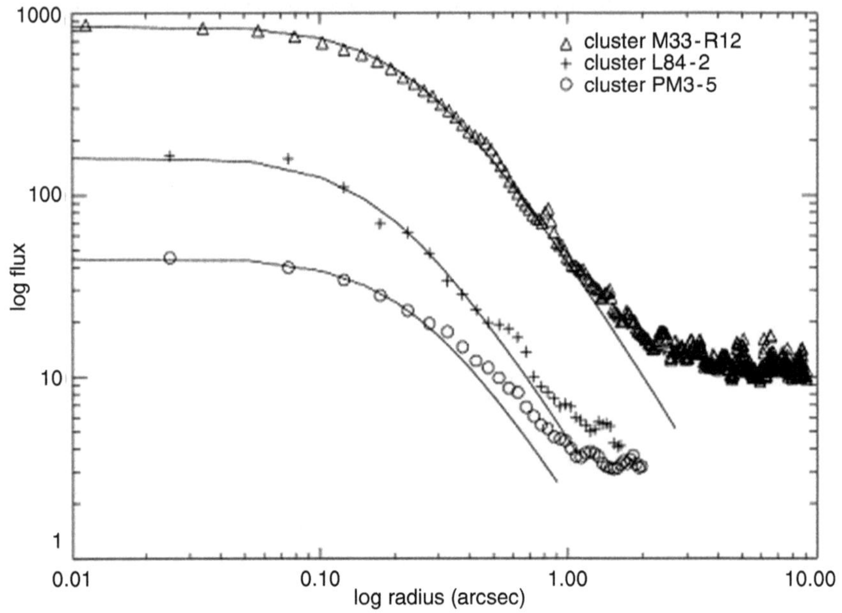

Fig. 6.13 Radial profiles of three M33 populous clusters (From Chandar et al. 1999b). Used by permission, copyright AAS

6.4 Very Young Clusters

To derive precise ages for very young clusters (<10^8 years) Chandar et al. (1999c) used fluxes measured with the HST at the far uv wavelength provided by the F170W filter. These data, combined with standard UBV data, were obtained for 15 clusters in M33. The F170 magnitudes ranged from 14.70 to 18.24 for the clusters detected with that filter, which were only 25% of the Chandar et al. (1999a) cluster sample. The addition of the uv fluxes provided improved measurements of the reddening and a better measure of the ages of these clusters, compared to results from UBV alone. Ages were found to range from 4×10^6 to 10^8 years and masses, derived by comparison with evolutionary models, ranged from 6×10^2 to 2×10^4 solar masses.

6.5 OB Associations

Like most other Sc spiral galaxies, M33 is notable for the rich population of stellar associations that outline its spiral arms. Conspicuous in the visible images (Sandage 1961) and, of course, even more in the ultraviolet, such as in the GALEX images (Thilker 2005), the OB associations have characteristics not notably different from those of the Milky Way Galaxy.

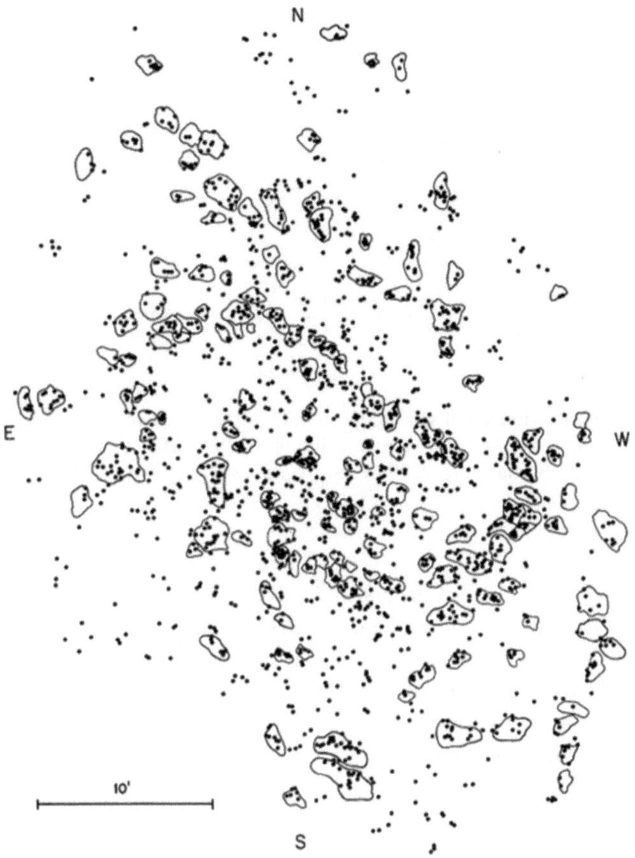

Fig. 6.14 A map of the OB associations in M33 identified by Humphreys and Sandage (1980). The OB associations are outlined and blue stars are indicated by *dots*. Used by permission, copyright AAS

The associations are generally referred to by the numbers assigned them by Humphreys and Sandage (1980). Their paper mapped and described apparent associations that were identified by eye inspection of Palomar Observatory 5-m telescope photographs (Fig. 6.14).

There have been several attempts to refine the Humphreys-Sandage catalog, most based on independent inspections by eye (e. g. Ivanov 1987). Some investigators interpreted the data to support the idea that there is a hierarchical distribution, with at least two levels, large associations made up of a collection of smaller groups (Efremov 1995). There have been several attempts to find a more unbiased method of selecting associations, using various statistical techniques (e. g., Battinelli 1991; and Wilson 1991, 1992). Most recently, evidence has been presented to support the idea that OB associations have no preferred size scale (Bastian et al. 2007, 2009).

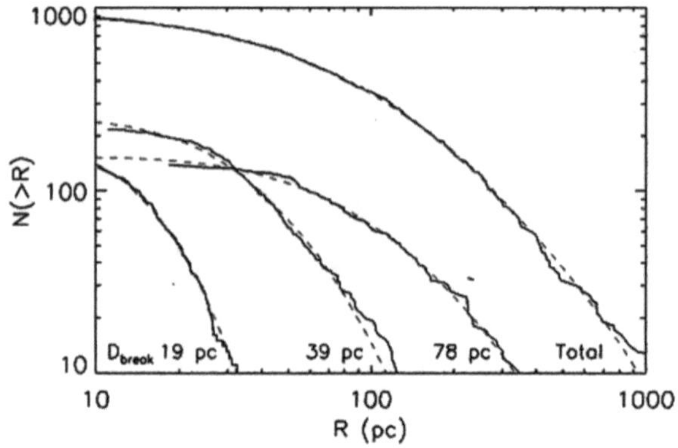

Fig. 6.15 The size distribution of star groups in M33, selected as a result of applying a minimum spanning tree method. The curves are for three different choices of D_{break}, a parameter related to the search criteria (From Bastian et al. 2009). Used by permission, copyright ESO

Using data for GMCs, star clusters and OB associations in M33, Bastian et al. (2009) found that for the entire range of scales, from 20 pc to several kpc, there is no preferred size, but a continuum of star-forming group sizes exists, the previous contrary results having been caused by spatial resolution and selection effects. The size distribution fits a lognormal distribution (Fig. 6.15) and the luminosity function fits a power law with a slope of –2.

6.6 Summing-Up

The galaxy M33 has a complete menu of different kinds of stellar groupings, from massive clusters and OB associations to small (~100 solar mass) clusters. It has disk clusters with disk-like kinematics and sub-solar abundances and it has halo clusters with halo kinematics. The abundances in the halo clusters tend to be higher in metals than in the Galactic globular clusters and the third parameter feature suggests that they include some intermediate age (~ 7–9 Gyr) clusters. Field stars show similar systematics, with evidence for a population of a wide range of age in the disk, a possible bulge or thick disk with an older, kinematically colder population and a very wide-spread halo. The picture emerges of a galaxy, less massive than the MW or M31, that differs from them in having had less chemical evolution and a slower initial episode of star formation.

There are many ways by which M33 can act as a good laboratory for study of star formation and cluster formation, due to its "nearly" face-on aspect, a feature where M31 and MWG are unfavorable. Among the many interesting global insights that have emerged is the conclusion that grouping of stars does not have a preferential size scale, a conclusion contrary to results published in decades of papers that found specific size scales for, by example, OB associations. This issue has been largely argued on either an empirical level or a statistical level. Of course, the more interesting level is the astrophysical one: what do the measured characteristics tell us about their formation and evolution?

References

Bastian, N., et al.: MNRAS **379**, 1302 (2007)
Bastian, N., Ercolano, B., Gieles, M.: A&SS **324**, 293 (2009)
Battinelli, P.: A&A **244**, 69 (1991)
Bedin, L.R., et al.: A&A **444**, 831 (2005)
Chandar, R., et al.: ApJS **122**, 431 (1999a)
Chandar, R., et al.: ApJ **517**, 668 (1999b)
Chandar, R., et al.: PASP **111**, 794 (1999c)
Chandar, R., et al.: A&A, **366**, 498 (2001)
Chandar, R., et al.: ApJ **646**, L107 (2006)
Christian, C.A., Schommer, R.A.: ApJS **49**, 405 (1982)
Cockcroft, R.: ApJ **730**, 112 (2011)
Efremov, Y.: AJ **110**, 2757 (1995)
Hiltner, W.A.: ApJ **131**, 163 (1960)
Humphreys, R.M., Sandage, A.R.: ApJS **44**, 319 (1980)
Huxor, A., et al.: ApJ **698**, L77 (2009)
Ivanov, G.: ApSS **136**, 113 (1987)
King, I.,: AJ, **71**, 64 (1966)
Lamers, H., et al.: A&A, **441**, 117L (2005)
Larsen, S.S., et al.: AJ **124**, 2615 (2002)
Melnick, J., D'Odorico, S.: A&AS **34**, 249 (1978)
Mochejska, B.J., et al.: Acta Ast. **48**, 455 (1998)
San Roman, I., et al.: ApJ **699**, 839 (2009)
Sandage, A.R.: The Hubble Atlas of Galaxies. Carnegie Inst, Washington (1961)
Sarajedini, A., Mancone, C.L.: AJ **134**, 447 (2007)
Sarajedini, A., et al.: ApJ **508L**, 37 (1998)
Sarajedini, A., et al.: AJ **120**, 2437 (2000)
Sarajedini, A., et al.: AJ **133**, 290 (2007)
Schommer., et al.: AJ **101**, 873 (1991)
Stonkute, R., et al.: AJ **135**, 1482 (2008)
Thilker, D.A.: ApJ **619**, L67 (2005)
Westerlund, B.: The Magellanic Clouds. CU Press, Cambridge (1997)
Wilson, C.D.: AJ **101**, 1663 (1991)
Wilson, C.D.: ApJ **384**, L29 (1992)
Zloczewski, K., Kalzuny, J., Hartman, J.: Acta Ast. **58**, 23 (2008)

Chapter 7
Gas and Dust

7.1 The General Interstellar Medium

The interstellar medium of M33 consists of five principal components: (1) neutral hydrogen gas, which is distributed broadly in a spiral structure, (2) high temperature excited gas that is primarily concentrated in luminous HII regions which are usually found in spiral arm segments, (3) very high temperature gas that is detected by x-ray telescopes, (4) interstellar dust, which can be detected either through reddening of stars or through infrared emission, such as mapped by the Spitzer Infrared Telescope, and (5) molecular gas, which can be detected by emission, especially by carbon monoxide molecules, and by absorption in the case of molecular hydrogen. Most of these topics are covered in this chapter.

7.2 HII Regions

M33 is rich in H II regions (Fig. 7.1). Early narrowband optical surveys revealed hundreds of them (Aller 1942; Courtés and Cruvellier 1965; Sabbadin et al. 1980; Courtés et al. 1987; and many others). A total of 1030 H II regions and 36 supernova remnants were identified prior to 1998 (Hodge et al. 2002). These objects, especially the brightest ones, have been the subject of a large number of papers, which have used them to study the composition, the abundance gradient, the star formation rate, the modes of star formation, and the velocity field of the galaxy.

7.2.1 The Population of HII Regions

The characteristics of M33's HII region population include its total number, its distribution of luminosities, its size distribution, its spatial distribution and the total Hα luminosity of the entire galaxy.

P. Hodge, *The Spiral Galaxy M33*, Astrophysics and Space Science Library 379,
DOI 10.1007/978-94-007-2025-1_7,

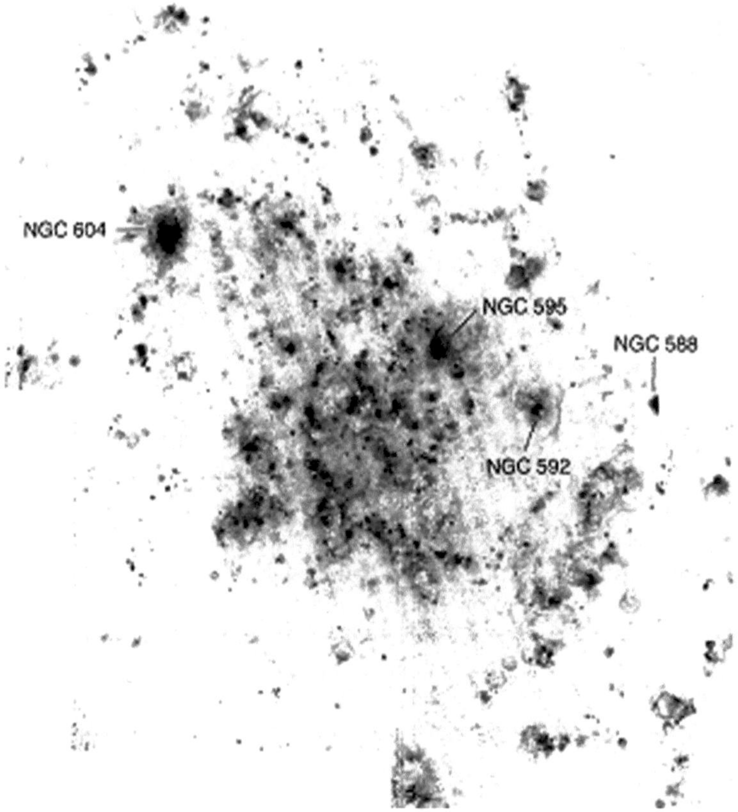

Fig. 7.1 An Hα image of M33 with the brightest HII regions identified (From Skelton 1999). Used by permission

The total number of HII regions in M33 is on the order of 3000; Hodge et al. (1999) cataloged and measured 2,338 emission regions, mostly Hα regions, and ~500 more were added by the Carlton College group (unpublished; see abstract in Blaha et al. 2009). The former survey was based on a mosaic formed by S. Gordon from four KPNO 4-m telescope images, while the latter was based on the Local Group Galaxy Survey (LGGS, Massey et al. 2006), which reached fainter limits (to a flux limit of 10^{-15} ergs-sec^{-1}-cm^{-2}) and covered a larger area.

7.2.2 The Hα Luminosity Function

The luminosity distribution of HII regions is a well-known measure of the HII region population in different galaxies. Differences in luminosity functions may reflect differences in the initial mass function of stars in different environments or it

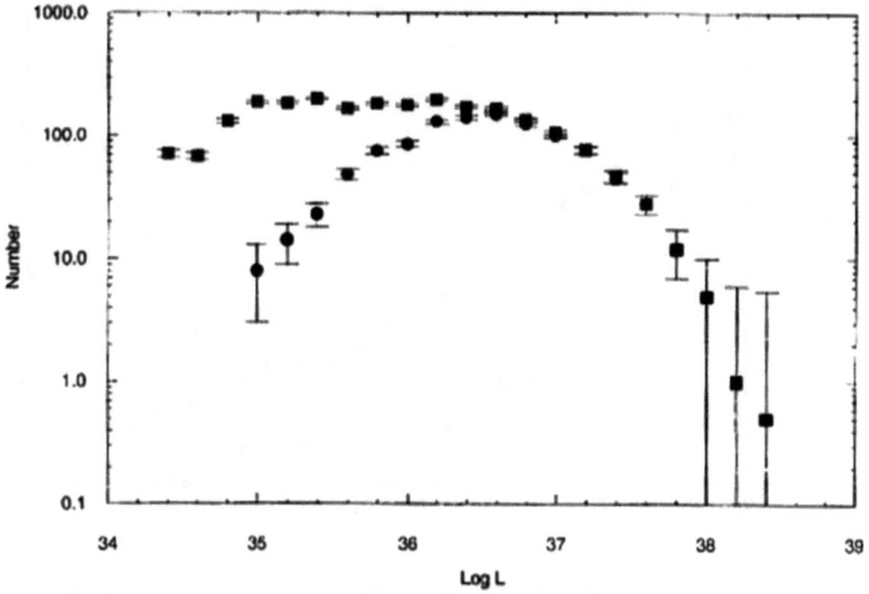

Fig. 7.2 The luminosity function for HII regions for the pre-1990 catalogs (*circles*) and the Washington sample (From Hodge et al. 1999). Used by permission, copyright AAS

may (in the case of density-bound HII regions) be governed by the characteristic density fluctuations of the interstellar gas in a galaxy. For M33 the luminosity function spans a wide range, from over 10^{38} to less than 10^{35} ergs sec^{-1}. The peak for the Washington data (Hodge et al. 1999) is at $L=10^{35.8}$ and the width is 2.4 dex (Fig. 7.2).

The high luminosity end of the curve is populated by massive complexes of hot stars and gas (see below) while the lowest-luminosity HII regions measured in M33 are either excited by single stars or are diffuse HII regions excited by the general uv field.

7.2.3 The Size Distribution

Another measure of the HII region population characteristics is the size distribution (van den Bergh 1981). For M33 the sizes were determined from isophotometry of KPNO 4-m telescope images (Hodge et al. 1999), which resolved emission regions down to ~5 pc in diameter. As found for distributions in other galaxies, the cumulative distribution is approximately exponential, with a characteristic scale parameter of 34+/−1 pc (Fig. 7.3). By comparison, the characteristic scale parameter of HII regions in the LMC is 86 pc and that for the Galaxy is 44 pc (van den Bergh 1981). It is perhaps not surprising that the diameter function would have a more or less

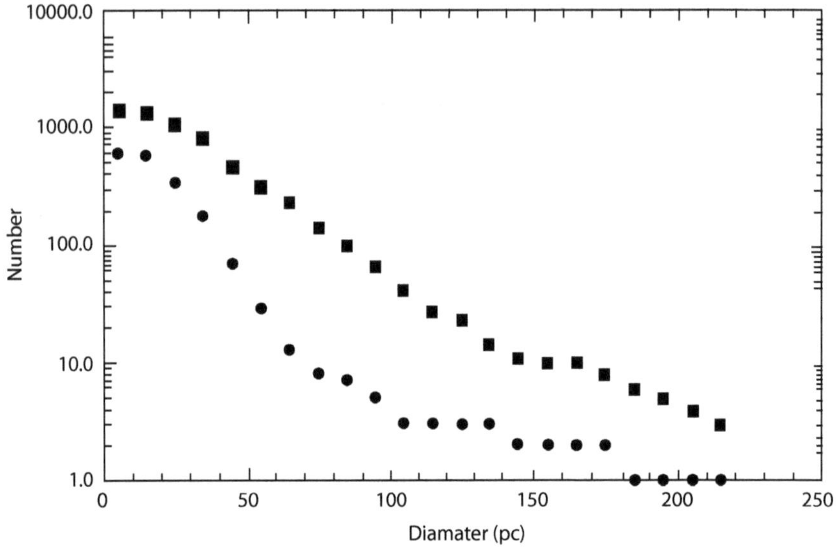

Fig. 7.3 The cumulative size distribution for HII regions in M33. The *upper curve* is for the complete data available in 1999 and the *lower curve* is for pre-1990 data (After Hodge et al. 1999). Used by permission, copyright AAS

universal shape, considering the fact that the sizes should be governed by the nature of the exciting stars and the gas, for which the astrophysics is similar, but the reason for the wide range in scale parameter is not clear.

The largest HII regions are over 100 pc in size, rivaling but not exceeding the size of 30 Doradus in the Large Magellanic Cloud, which is the largest nearby example of an HII region in a normal galaxy (e.g., not in a super-luminous interacting or merging environment).

7.2.4 The Spatial Distribution

The distribution of HII regions in the disk of M33 can be determined from the various catalogs. However, most searches were limited in extent, with the outer portions of the disk not represented. For the inner disk, it is possible to use earlier results, which extend to ~23 arcmin. Figure 7.4 shows the rectified radial distribution (based on the geometry of the disk given in Chap. 1). Although there is large scatter, the data fit an exponential fairly well:

$$\log N = -8e^{-5R} + 2.18$$

The scale length is similar to that for blue light (Chap. 4).

The total Ha luminosity of M33 has been calculated to be 2.8×10^{40} ergs s^{-1} (Devereux et al. 1997).

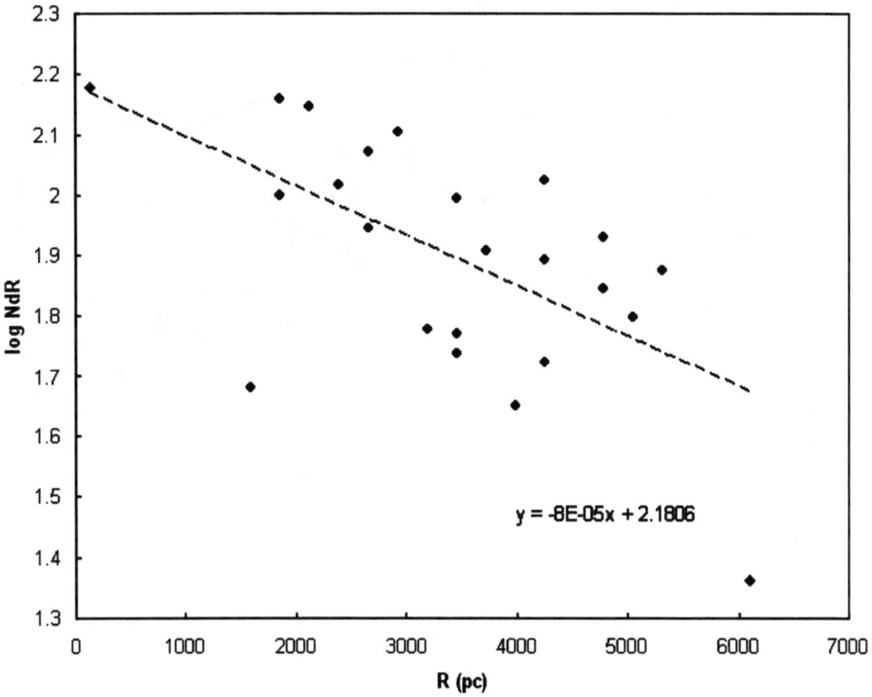

Fig. 7.4 The radial distribution of HII regions in the inner M33 disk

7.2.5 Massive Star Content

Many recent papers have dealt with the stellar content of HII regions and their environs. The two primary scientific goals behind these studies are the question of the energy balance between the nebula and the exciting stars, which gives information on the structure and physical conditions of the gas, and the shape of the massive end of the IMF, which is best determined in very young, massive star groups.

Hunter (1994) and Hunter et al. (1995) carried out a nice study of the massive stars in HII regions, concentrating on shell-like nebulae, where the geometry is relatively simple. From the characteristics of the embedded stars, they calculated that the flux of uv light is larger by a factor of 4–10 than that required to ionize the nebulae. This implies that a large amount of the stellar uv spreads out beyond the gas shells into interstellar space. Simple models of formation of the shells are found to be inadequate to completely explain the structure, implying a complicated history of interactions between stars and gas (Fig. 7.5).

On the other hand, Oey and Massey (1994) selected two giant shell nebulae in M33 to identify the exciting stars. Finding a dominant single O9 star at the centers, they compared the characteristics of the stars and nebula with models of wind-blown shells and found good agreement, when they took into account the evolution of the stars and their winds.

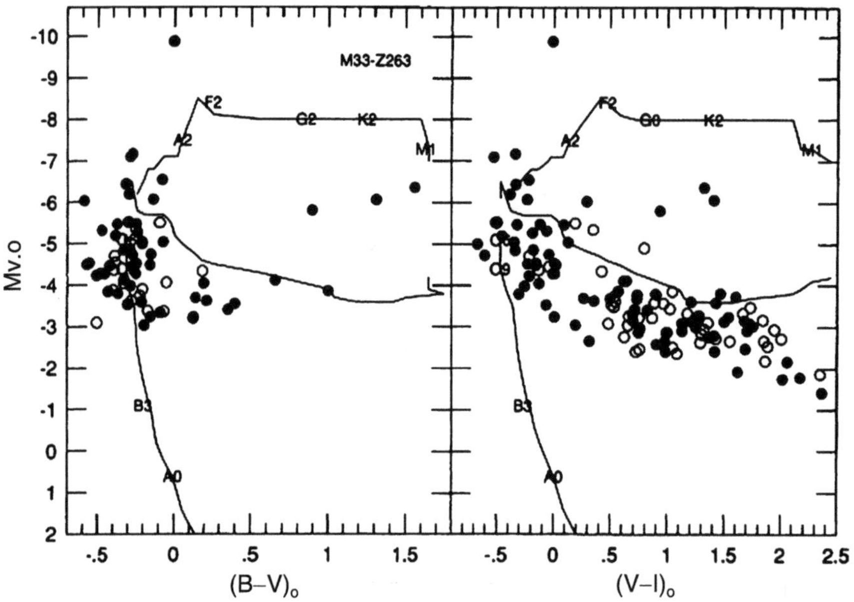

Fig. 7.5 The color-magnitude diagrams of the embedded stars in HII region Z263 as measured by Hunter et al. (1995). The *lines* define the main sequence and the Ia and Ib branches. A few spectral types are indicated. Used by permission, copyright AAS

Skelton et al. (1999) carried out a comprehensive study of seven of M33's giant HII regions, obtaining observations from the ground, from HST and from the ISO observatory. She modeled the structure, content and characteristics of the gas, stars and dust. Details, however, have not been published except in her PhD thesis.

Jiang et al. (2002) used the Beijing Observatory Schmidt telescope to measure the spectral energy distributions for 145 HII regions selected from the Boulesteix et al. catalog (1974). They found that the typical stellar content of the HII regions consists of young clusters with the combined luminosity of a main sequence O5 star.

Pellerin and Drissen (2005) used far-ultraviolet measurements of giant HII regions in M33 to derive ages, stellar masses, the stellar IMF and the Hα and V fluxes. Only for fairly massive objects, with 10^3 solar masses in massive stars, could the stellar synthesis method provide accurate data, because of small number statistics effects.

7.2.6 Atomic Abundances

Mayall and Aller (1942) observed 25 HII regions distributed over the main disk (for r < 30 arcmin) with the primary aim of measuring the rotational velocity curve.

They used low dispersion spectra (304 A/mm in the blue) and were able to determine radial velocities and a rotation curve for the galaxy (Chap. 12). Aller (1942) used the spectra to make a preliminary assessment of the excitation level of the gas, based on the ratio of visually-estimated strengths of the 5007/3727 Å emission lines.

Garnett et al. (1992) observed the HII region MA 1 both with spectra and Hα images, finding the [O II] lines to show a radial gradient out from the central star(s), from quite weak near the center to normal near the edge. The temperature was found to be approximately 14,000 K and the implied oxygen abundance is 1/10th the solar value.

Han et al. (2001) published a spectroscopic survey of HII regions in M33, determining the abundances of O, S and N. They found that the abundances of these elements are low with respect to H, compared to the MWG or M31. A radial dependence was detected, as found by others.

Crockett et al. (2006) carried out a study of 13 HII regions in M33. They found that they could detect the [O III] λ4363 line in six of them, allowing the electron temperature to be measured directly from the line ratios. The derived gradients are –0.016 (+/–0.017) dex kpc^{-1} for N and –0.012 (+/–0.011) dex kpc^{-1} for O/H. These agree well with predictions from nucleosynthesis theory.

7.2.7 NGC 604

The most massive HII region in M33 is the giant object NGC 604, larger and more luminous than any HII region in the Local Group except for 30 Doradus in the LMC. Multiple papers on NGC 604 have studied its characteristics, using the whole array of wavelengths, including optical (HST), radio (VLA), infrared (Spitzer), UV (Galex), and X ray (Chandra) observations (Fig. 7.6).

As an index of the giant star content of NGC 604, a Hubble Space Telescope search found 14 Wolf-Rayet stars. Such a large number of massive stars is an indication of the large amount of energy and mechanical power involved in the H II region. The Hα images show many ringlike features which are identified with some of the highly mass, high temperature stars involved in star formation.

The hot ionized gas in NGC 604 has been studied using the Chandra space telescope (Tullman et al. 2008). X-ray emission was detected from most of the optical extent of the H II region. The total x-ray luminosity is measured to be 9×10^{35} erg/s. Most of the x-ray emission comes from inside the bubbles that are seen in the optical image. The emission can be explained if there are approximately 200 OB stars in the complex. Evidence for the presence of supernova remnants is also found in the outer parts of NGC 604. Spitzer Space Telescope observations of NGC 604 show that the cool medium has a temperature between 35 and 49°Kelvin. Unlike the smaller HII regions, NGC 604 has nearly half of its gas residing in photodissociation regions, rather than nearly all in HII regions (Fig. 7.7).

Fig. 7.6 The giant HII region NGC 604, shown here in an HST *color version* (NASA)

7.3 Molecular Gas

M33 has been surveyed for molecular clouds in recent years, for instance, by Wilson and Scoville (1990). For the 38 giant molecular clouds detected by Wilson, the characteristics were found to be similar to the population found in our Galaxy. In particular the ratio of carbon monoxide flux to the column density of molecular hydrogen is nearly the same as for our Galaxy and the distribution of masses of the giant molecular clouds is similar to the local sample, except for a cutoff for masses larger than 400,000 solar masses.

A detailed mapping of the molecular clouds of M33 was published by Engargiola et al. (2003), using the BIMA interferometer, giving a resolution of approximately 50 pc. The mass limit of the survey is approximately 150,000 solar masses. A total of 158 giant molecular clouds was detected, comprising a total mass of 23 million solar masses. The authors estimate that about half of the total carbon monoxide flux is included in their survey, implying a total mass of 45 million solar masses for the galaxy. The distribution of clouds correlates with the positions of dense lanes of

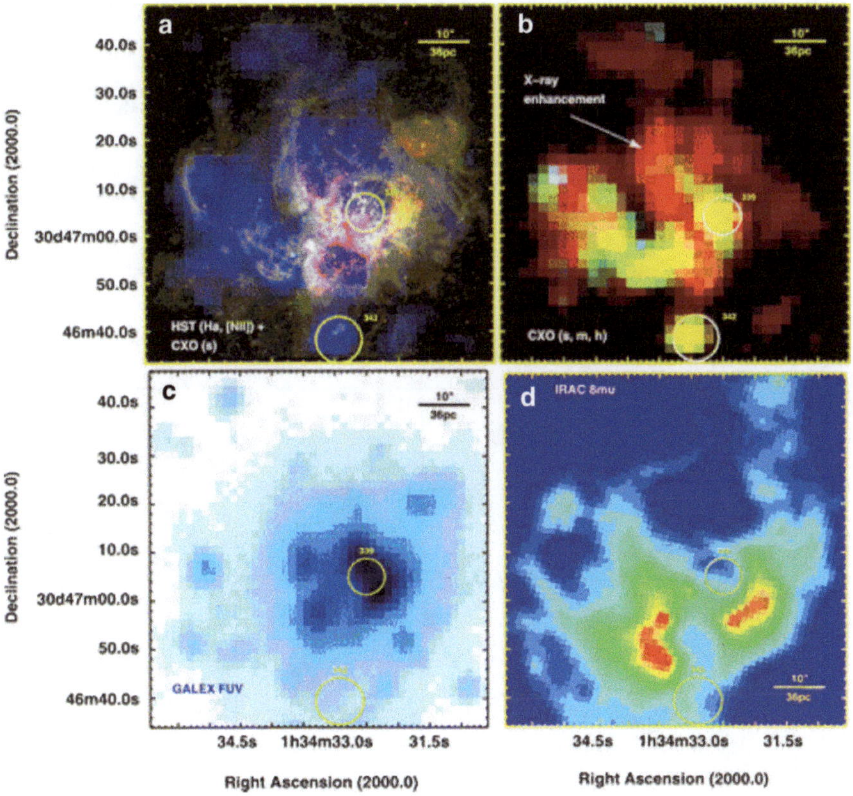

Fig. 7.7 Four images of NGC 604. The *upper left* image shows the emission coded by color from: *red*: Hα, *green*:][NII] and *blue*: soft x-rays. The *upper right* is an x-ray image, with the *red* indicating the soft, *green*: the medium, and *blue*: the hard band. The *lower left* image is a far ultraviolet image, showing the abundance of very hot stars, and the *lower right* image is an infrared map from the 8 μm survey. *Yellow circles* are discrete x-ray sources (From Tullmann et al. 2008). Used by permission, copyright AAS

H I and with luminous H II regions. This implies that the giant molecular clouds form rapidly from atomic gas clouds and have only a short period before massive stars begin to form (Fig. 7.8).

In 2007 new high resolution observations of CO were obtained using the Nobeyama Radio Observatory 45 m telescope (Rosolowsky et al. 2007). In addition to a new catalog of clouds this study showed that the fraction of gas found in giant molecular clouds is different in different parts of the galaxy. Near the center 60% of the gas is in clouds whereas at a distance of 4000 pc from the center that amount is reduced to 20%. The CO flux density from the bright arms of M33 is approximately 50% higher than from the interarm regions. The masses of giant molecular clouds are found to fit a power law distribution with index of −2.1.

Using the Spitzer infrared observatory, astronomers have been able to measure lines of many species, including forbidden lines of [S IV] and [N II] in many H II regions.

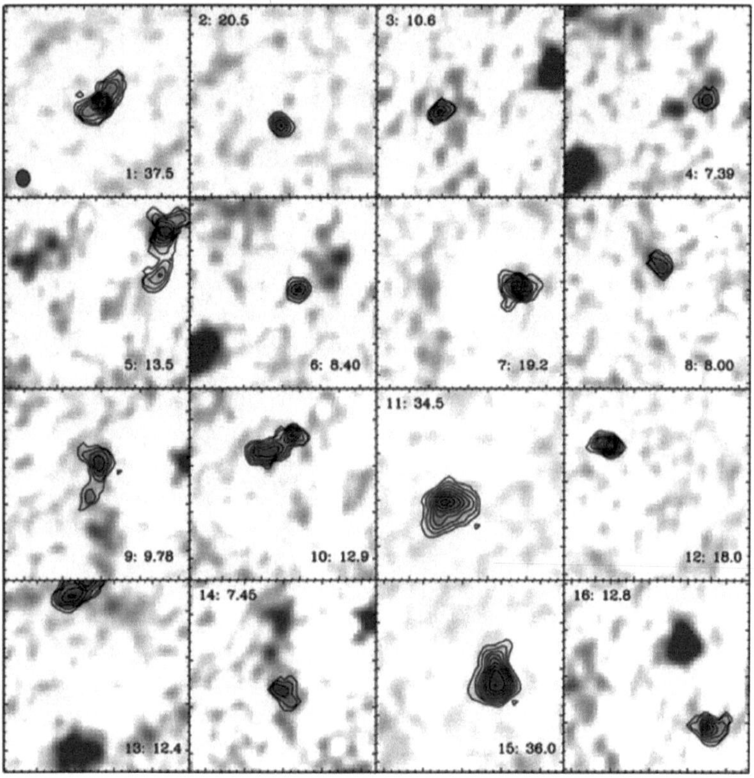

Fig. 7.8 Examples of giant molecular clouds in M33, as observed with the BIMA interferometer. The *gray scale* shows the CO brightness integrated over velocity, while the *lines* are contours of the targeted cloud (From Rosolowsky et al. 2003). Used by permission, copyright AAS

Using these spectra it was found that the ratio of neon to sulfur does not vary as a function of distance from the center of the galaxy. Using the James Clerk Maxwell telescope it was found that the molecular content of the H II regions in M33 is very similar to that in the Galaxy's molecular clouds.

Study of the molecular clouds in M33 by Wilson et al. (1997) showed that the carbon monoxide temperature depends on the location of the giant molecular cloud. The temperature is between 10° and 20° Kelvin outside of the H II regions, and it is 15–100° Kelvin inside the H II regions. It is at least 100° Kelvin in the giant HII region NGC 604. It was proposed that the unusual properties observed in that region are the result of its extended period of star-forming activity (Fig. 7.9).

High-resolution CO observations were obtained by Gratier et al. (2010) with the 30-m IRAM Telescope. The resolution was 12 arc sec × 2.6 km/s and the coverage was of the main body of the galaxy, out to 8 kpc, chiefly along the major axis. The total luminosity for the galaxy was estimated by scaling to the full disk, giving 2.8×10^7 K k/s/pc^2. This corresponds to 3.3×10^8 solar masses of H and He. The CO

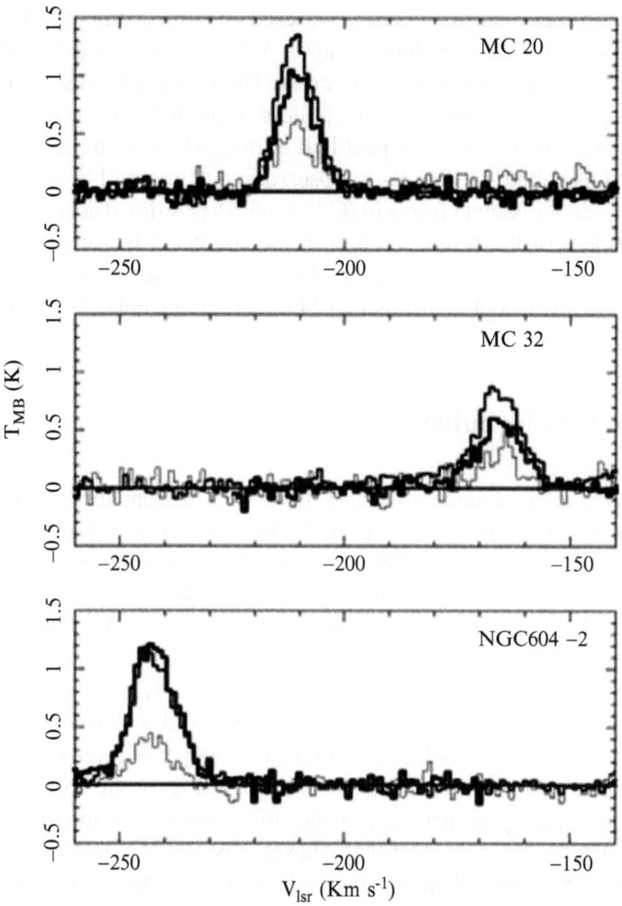

Fig. 7.9 Spectra of three giant molecular clouds in M33. The molecules are ^{12}CO J=2_1, ^{13}CO J=2_1, and ^{12}CO J=3_2. The ^{13}CO curve has been increased by a factor 3 (From Wilson et al. 1997). Used by permission, copyright AAS

surface brightness does not follow the nearly level HI surface brightness, but instead, decreases outward with a scale length of –1.9 kpc, similar to that of the disk stars.

The interstellar hydrogen molecule in M33 has been detected in the most luminous H II regions, though the detection is extremely difficult because of the very faint absorption. Recent detections were the result of observations made by the FUSE space telescope, which examined the four brightest HII regions in M33 and found absorption lines identified with molecular hydrogen in three of them, NGC 588, NGC 592, and NGC 595 (Bluhm et al. 2003). Column densities along the line of sight for these detections were between 10^{14} and $10^{17}/cm^2$. The presence of molecular hydrogen was not unexpected, but the experiment was important because heretofore this molecule has been virtually undetectable from the ground.

Water vapor masers in M33 were discovered in 1990. Using the VLBI, Greenhill et al. (1990) constructed a synthesis map of M33, in which was found ten masers, located in an area 300 milliarcseconds across. There were 14 individual components and the positions were measured accurately enough to be used for astrometry. The objects are found near a compact H II region, which is in the complex of star-forming regions called IC 133. Later observations (Greenhill et al. 1993) revealed an additional set of distinct masers in IC 133, with a positional accuracy that allowed the measurement of the proper motion of the sources. Five of the maser features gave accuracies of position averaging about 10 microarcseconds. The values are consistent with the accepted distance of M33 based on Cepheid variables.

7.4 Planetary Nebulae

Planetary nebulae are abundant in the galaxy M33. Recent surveys include that of Magrini et al. (2000), who used the Isaac Newton telescope in La Palma, Spain. The survey resulted in a catalog of 145 candidate planetary nebulae. The data allowed the derivation of the luminosity function, which in other galaxies has been found to provide an estimate of the distance. In the case of M33 the derived distance is 840 kpc, which is reasonably close to the usually quoted distance (see Chap. 1).

A more complete survey of the planetary nebula system was carried out by Ciardullo et al. (2004). Using both photometry and spectroscopy the authors identified 152 planetary nebulae, spanning a total range of three magnitudes. As in the case described above, the distance was derived from the apparent magnitude of the brightest planetary nebulae, at the brightness of an apparent sharp cutoff (Fig. 7.10). Unlike the case described above, however, the answer is a distance of 940 kpc, fully 15% larger than the standard distance. Rather than implying an error in the distance, this result is probably accounted for by a different assumption regarding extinction.

Radial velocities determined from spectroscopy of 140 planetary nebulae allow the conclusion that the planetary nebulae are kinematic members of the old disk population (Ciardullo et al. 2004). The radial velocity distribution is nearly uniform over the entire galaxy. The authors state that their data argue against mass models that include dark matter that is limited to the plane of the galaxy.

7.5 Supernova Remnants

Several methods have been used to detect supernova remnants (SNRs) in M33. Optical methods use the ratio of Hα to [SII], either through spectroscopy or by using narrowband images of the galaxy. Radio observations can use radio spectra to identify the nonthermal sources in M33. X-ray observations are able to detect those supernova remnants that have very high temperature gas emitting sufficiently in the

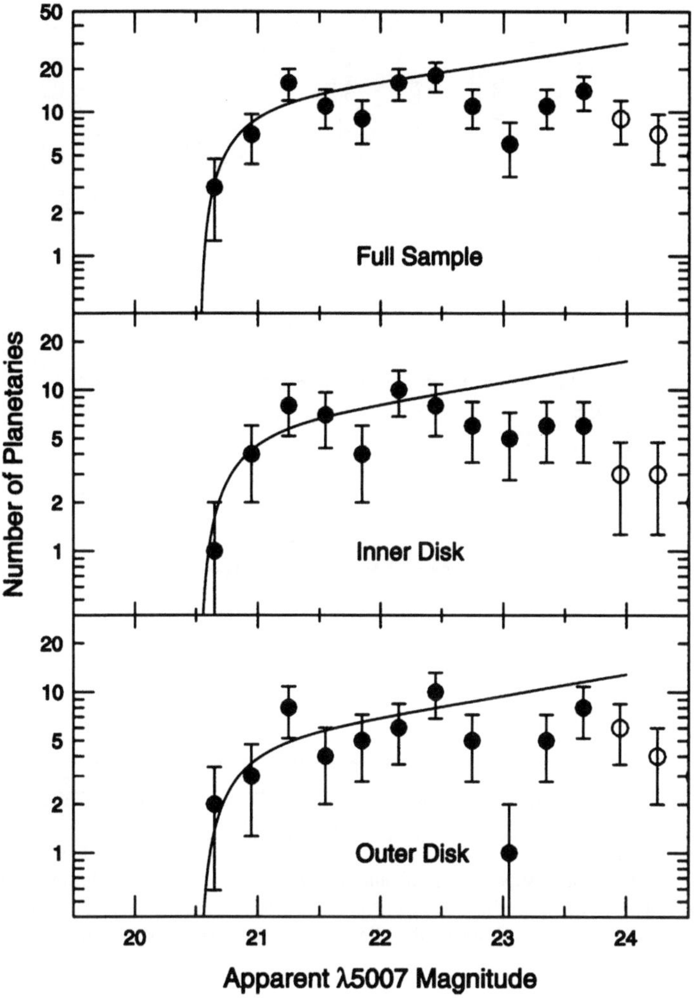

Fig. 7.10 The luminosity function for M33 planetary nebulae (From Ciardullo et al. 2004). Used by permission, copyright AAS

x-ray spectral region for detection. Finally, optical characterization of the supernova remnant candidates can be made with high-resolution images, such as taken by the Hubble Space Telescope.

Among the early studies of M33's supernova remnants was a survey using optical spectroscopy (Smith et al. 1993). Optical spectra were obtained of 42 emission nebulae, all suspected supernova remnants in the inner portion of M33. All of the candidates were found to have a brightness ratio of Hα to forbidden ionized sulfur ([SII]) of 1.6 or smaller, a characterization of supernova remnants. The spectra of many of the candidates also showed emission of forbidden neutral oxygen ([OI]),

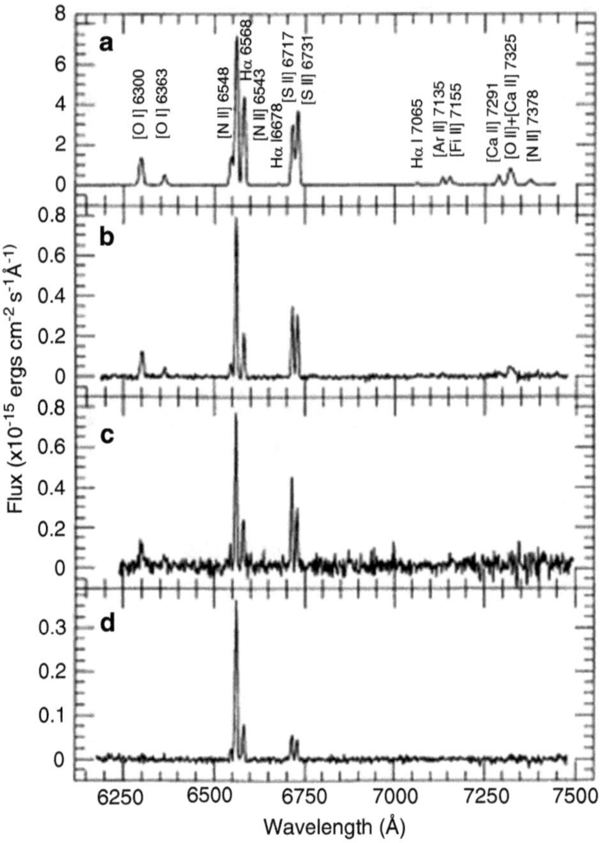

Fig. 7.11 Spectra of three supernova remnants. The *upper* is one of the most luminous in M33, the *next down* is an average case, the *third* is one of the faintest and the *bottom* spectrum is of an HII region for comparison (From Smith et al. 1993). Used by permission, copyright AAS

which is also characteristic of SNR's. Also the strength of the lines provided values for the oxygen abundance as a function of distance from the center of M33, which agreed with the gradient found from HII regions (Fig. 7.11).

The structure of the supernova remnants in M33 was examined using the WFPC2 on the Hubble Space Telescope (Blair and Davidsen 1993). The images clearly showed the kinds of structures well-known for supernova remnants in, for example, the Milky Way and the Magellanic Clouds. The objects showed rings, partial arcs and discs. High dispersion spectra obtained with the Kitt Peak Observatory 4 m telescope showed broad emission lines with widths between 165 and 275 km/s. These high velocities indicate the presence of the rapid expansion of the observed filaments and shells.

An early survey of X-radiation from M33 was carried out with the ROSAT Space Observatory, which was soon followed by the more sensitive XMM-Newton survey of M33 (Pietsch et al. 2004). The sources detected included supernova remnants as well as x-ray binaries and various background objects. Using x-ray spectral energy

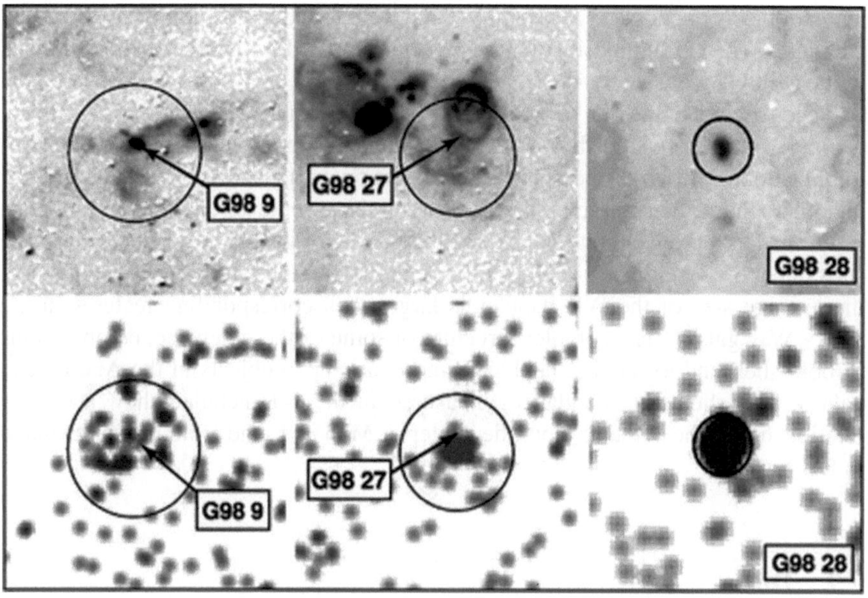

Fig. 7.12 Soft band (0.35–1.1 keV) Chandra images of three M33 SNRs (*bottom panels*), compared to H-alpha images (*upper panels*) (From Ghavamian et al. 2005). Used by permission, copyright AAS

distributions combined with optical and radio emission, it was possible to distinguish between different kinds of sources. There were 408 detected sources of which 44 were either confirmed supernova remnants or strong candidates. Among the remaining sources were five supersoft sources, two x-ray binaries and 267 additional sources that were classified as hard, and which may include crab-like SNRs. Other sources were found to be foreground stars or distant active galactic nuclei.

The Chandra X-ray Observatory allowed an even more sensitive survey of M33 (Ghavamian et al. 2005). Combining the Chandra images with optical and radio indicators revealed a total of 100 supernova remnants that are optically emitting detectable radiation, of which 37 are x-ray emitting as well (Fig. 7.12). The sample was compared to the supernova remnants with luminosities in excess of 10^{35} ergs/s in the Large Magellanic Cloud with the result that M33's SNR's are only half as common as in the LMC. This difference has not been completely explained; it is unlikely to be due to differences in, for instance, the abundance differences of those two galaxies. For more details about X-ray sources, see Chap. 11.

7.6 WR Nebulae

The galaxy M33 has been found to have many emission line stars, many of which turned out to be Wolf-Rayet stars (Massey et al. 2007). Drissen et al. (1991) identified 11 WR nebulae and found that the nebulae were systematically larger than similar objects in our galaxy. Several of them were found in the form of rings or shell structures. The Hα luminosity of these objects compared well with windblown shells in the Milky Way.

Spectroscopy of the ring nebulae around seven WR stars was obtained by Esteban et al. (1994). The spectra showed that the nebulae were primarily powered by photoionization. For two of the nebulae there was no evidence for expansion, and for the others the velocity structures were found to be complex. For some the evidence suggests that there is an internal expanding bubble of gas with the whole structure surrounded by a stationary shell. As found in the previous study mentioned above, the average size of the WR bubbles is larger than corresponding features in the Milky Way galaxy. The complex structure of some of these nebulae, combinations of both shell and expanding bubble, is more similar to objects in the Andromeda galaxy. The chemical abundances measured in the WR nebulae show a similar relationship to the distance from the center of M33 as found from measures made in HII regions.

7.7 Dust

The IRAS Space Telescope provided the first high resolution mapping of the infrared emission from M33 (Deul 1989). The four different ir wavelengths (12, 25, 60 and 100 μm) allowed the separation of sources into two components, a point source component and a diffuse component. The 12 and 25-μm radiation comes primarily from a dusty disk, made largely of small particles. The point sources are bright at the longer wavelengths (Fig. 7.13).

A new mapping of the dust content of M33 was carried out in 2002 using the European Space Agency's Infrared Space Observatory (Hippelein et al. 2003). The German camera ISOPHOT was used to map the galaxy at three wavelengths, 60, 100, and 170 μ. The structure of the dust was shown in some detail, reflecting the spiral structure and pinpointing the position of several star-forming areas. A total of 60 HII regions showed enhanced infrared emission and the ratio of far infrared flux to Hα emission was found to be correlated with distance from the center of M33, in conformance with the decreasing extinction outwards.

These measurements confirmed the existence of two dust components in the disk of M31, a warm and a cold component. The cold component is widely distributed over the disk, apparently warmed by the diffuse interstellar radiation. The warm component is mostly detected in the central area and in the spiral arms and is apparently heated by the high temperature stars in star-forming regions (Fig. 7.14).

In 2008 further observations of M33 were carried out by the Spitzer space telescope (Rubin et al. 2008). These observations showed in remarkable detail what M33 looks like at long wavelengths. The images revealed giant star-forming regions, dust shells surrounding stars, HII regions and many red supergiant stars, with their dusty envelopes. Probably the most important results had to do with observations of very young dust-embedded star clusters. These objects had been detected previously

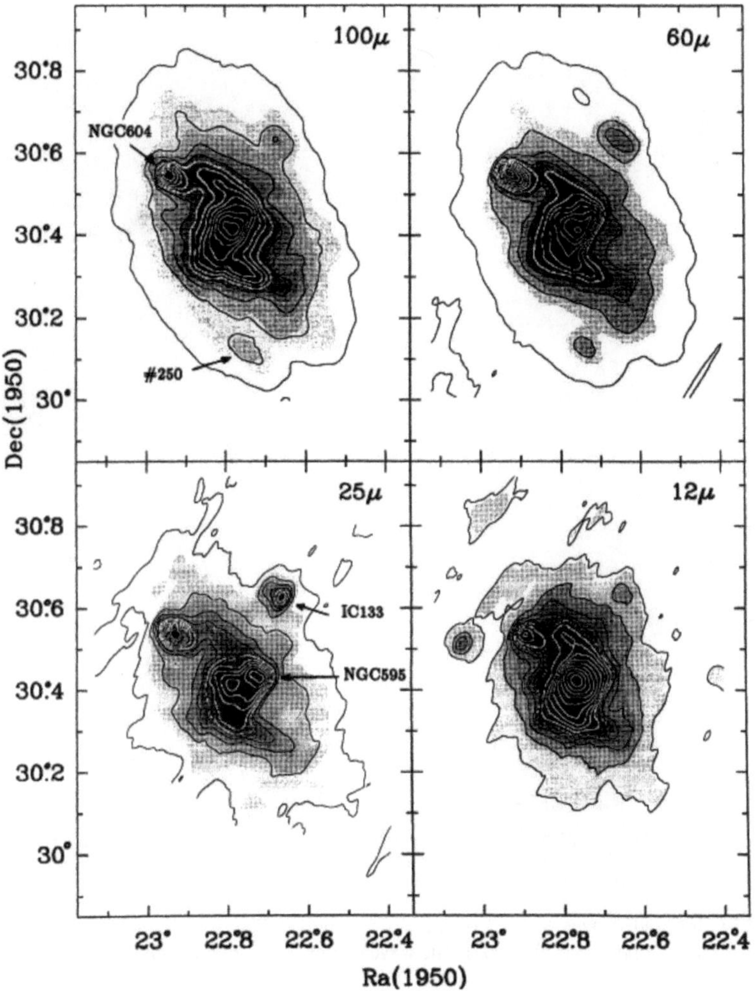

Fig. 7.13 Maps of the infrared radiation from the IRAS satellite at four different wavelengths (From Deul 1989). Used by permission, copyright ESO

using radio telescopes but their precise nature had not been determined because they were invisible at optical wavelengths. However in the Spitzer images they are detectable at a range of infrared wavelengths from 3.6 to 160 μ. Their spectral energy distributions allow a determination of their physical properties and the nature and amount of the dust that obscures them. The clusters were found to be very young, about one million years old, and they are associated with shell structures, core-halo structures, cometary tails and irregular dust morphologies.

A number of astronomers have taken advantage of the Spitzer and GALEX Space Telescopes to explore the diffuse content of galaxies as seen from the extremes of

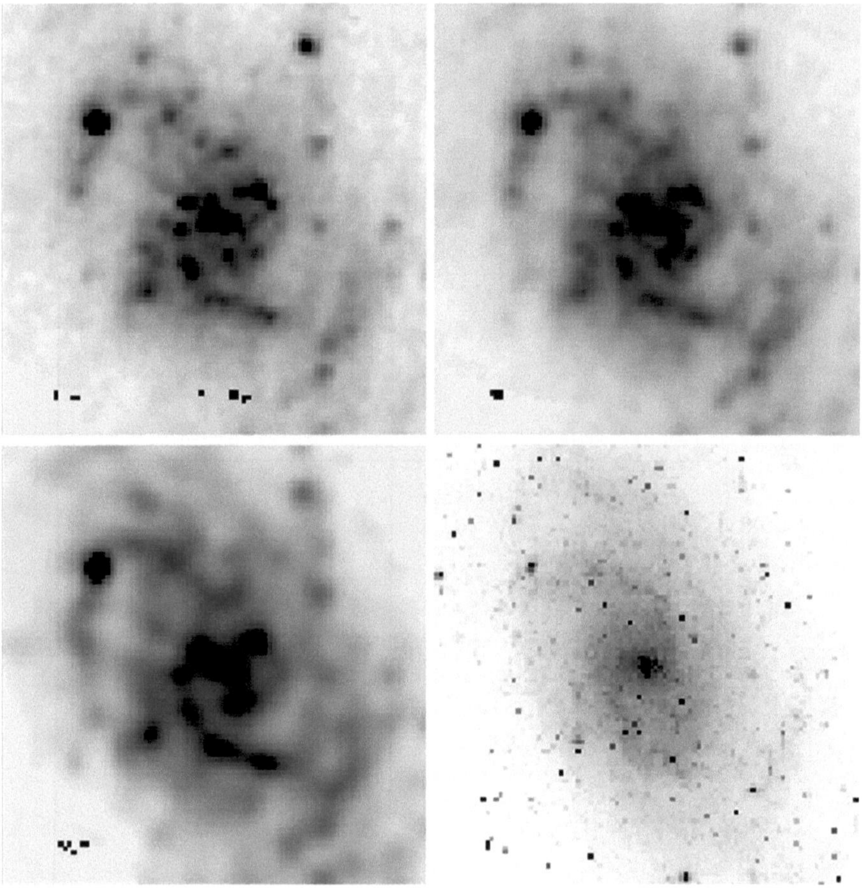

Fig. 7.14 Infrared images of M33 obtained from the ISOPHOT instrument on the ISO space telescope (from Hippelein et al., 2003). The wavelengths in microns are *top left*: 60, *top right*: 100, *bottom left*: 170 and *bottom right*: visual R. Compare with Fig. 7.13 to see the improvement of resolution. Used by permission, copyright ESO

wavelength that those space telescopes provided. For example, Verley et al. (2010) have combined the uv, optical and ir data for M33 to analyze its global interstellar content. After separating out discrete sources, at wavelengths of 8 and 24 μm, the diffuse component is 60% of the emission in the central 3 kpc, while this percentage increases outwards to a value of 80%. The total extinction, which has a mean value of $A_V = 0.25$ magnitudes, decreases outwards, following the shallow decrease of metallicity. The star formation rate is calculated to be 0.45 (+/−0.10) solar masses per year for the recent 100 Myr period.

7.8 Summing-Up

The interstellar material in M33 consists of the same ingredients as found in our Galaxy or M31, but with differences, some of which can be ascribed to its smaller mass. Chiefly among these differences are the chemical abundances. The ratio of heavier elements to hydrogen in the HII regions is approximately one-tenth that in the Milky Way. As in the Galaxy, there is a shallow gradient in the heavy element abundances with distance from the center, on the order of –0.01 dex/kpc. The HI and the HII regions are both concentrated to the spiral arms, but these arms are not well-developed in shape. The HI is warped with respect to the plane of the stellar disk, as in the Milky Way. There are nearly 3000 HII regions in the catalogs and they range in H-alpha luminosity up to 10^{38} ergs/s. Several giant HII regions are found that have complicated structure, showing the effects of multiple star-forming events and supernova activity.

Molecular clouds are prominent in the disk and they show a distribution of masses similar to the Galaxy's. Based on measurements of CO and H_2, the total mass of molecular gas is calculated to be approximately 3×10^8 solar masses. Forms of interstellar gas related to late stages of stellar evolution are the planetary nebulae and supernova remnants. More than 100 of each have been studied and their properties are found to be similar to those in the MWG.

The dust in M33 has been the subject of much recent research, especially by infrared measurements using space telescopes. There are basically two components of the infrared sources, warm point sources that are related to luminous HII regions and a diffuse cold component that is widely distributed over the disk and is heated by the diffuse interstellar radiation.

References

Aller, L.H.: ApJ **95**, 52 (1942)
Blaha, C., et al.: BAAS **38**, 939 (2009)
Blair, W.P., Davidsen, A.F.: PASP **105**, 494 (1993)
Bluhm, H. et al.: A&A **398**, 983 (2003)
Ciardullo, R., et al.: ApJ **614**, 167 (2004)
Courtes, G., Cruvellier, P.: Ann Ap **28**, 683 (1965)
Courtes, G., et al.: A&A **174**, 28 (1987)
Crockett, N.R., et al.: ApJ **637**, 741 (2006)
Deul, E.R.: A&A **219**, 78 (1989)
Devereux, N., et al.: AJ **113**, 236 (1997)
Drissen, L., Shara, M.M., Moffat, A.F.J.: AJ **101**, 1659 (1991)
Engargiola, G., et al.: ApJS **149**, 343 (2003)
Esteban, C., Vilchez, J.M., Smith, L.J.: AJ **107**, 1041 (1994)
Garnett, D.R., Odewahn, S.C., Skillman, E.D.: AJ **104**, 1714 (1992)
Ghavamian, P., et al.: AJ **130**, 539 (2005)
Gratier, P., et al.: A&A **522**, 3 (2010)

Greenhill, L.J., et al.: ApJ **364**, 513 (1990)
Greenhill, L.J., et al.: ApJ **406**, 482 (1993)
Han, S.R., et al.: J. Korean Ast. Soc. **34**, 67 (2001)
Hippelein, H., et al.: A&A **407**, 137 (2003)
Hodge, P. et al.: PASP **111**, 685 (1999)
Hodge, P.W., Skelton, B., Ashizawa, J.: An Atlas of Local Group Galaxies. Astrophysics and Space Science Library, Vol. 221. Kluwer, Dordrecht (2002)
Hunter, D.A.: AJ **108**, 1658 (1994)
Hunter, D.A., Boyd, D.M., Hawley, W.N.: ApJS **99**, 551 (1995)
Jiang, H., et al.: AJ **124**, 3179 (2002)
Magrini, L., et al.: A&A **355**, 713 (2000)
Massey, P., et al.: AJ **131**, 2478 (2006)
Massey, P. et al.: AJ, **134**, 2474
Mayall, N.U., Aller, L.H.: ApJ **95**, 5 (1942)
Oey, M.S., Massey, P.: ApJ **425**, 635 (1994)
Pellerin, A., Drissen, L.: BAAS **37**, 1271 (2005)
Pietsch, W., et al.: A&A **426**, 11 (2004)
Rosolowsky, E., et al.: ApJ **599**, 258 (2003)
Rosolowsky, E., et al.: ApJ **661**, 830 (2007)
Rubin, R.H., et al.: MNRAS **387**, 45 (2008)
Sabbadin, F., Rafinelli, P., Biachini, A.: A&AS **39**, 97 (1980)
Skelton, B.: BAAS **31**, 1509 (1999)
Smith, R.C., et al.: ApJ **407**, 564 (1993)
Tullmann, R., et al.: ApJ **85**, 919 (2008)
Tullmann, R., et al.: ApJS **193**, 31 (2011)
van den Bergh, S.: AJ **86**, 1464 (1981)
Verley, S., et al.: A&A **518**, 68 (2010)
Wilson, C.D., Scoville, N.: ApJ **363**, 435 (1990)
Wilson, C.D., Walker, C.E., Thornley, M.D.: ApJ **483**, 210 (1997)

Chapter 8
Spectra and Abundances

One of the important characteristics of individual galaxies is the abundance of the different chemical elements. In general, galaxies of large mass, such as our Galaxy and the Andromeda galaxy, have abundances of the heavier elements that are high, while low mass galaxies, such as the Magellanic Clouds and dwarf galaxies in the Local Group, have abundances that are ten or more times smaller. The chemical abundances in galaxies are frequently referred to as "metallicity", because for stars the most conspicuous absorption lines are of iron and other metals. The lighter elements are usually measured in emission nebulae, where emission lines of hydrogen, oxygen, and other light elements can be measured. In this chapter we take up, in turn, spectroscopic measurements of stars in M33, stellar photometric evidence of abundances, spectroscopic evidence for abundances determined for emission regions, and evidence for the existence of the relationship between abundances and distance from the center of M33, the "metallicity gradient".

8.1 Stellar Spectroscopy

The first stellar spectroscopy of stars in M33 concentrated, for obvious reasons, on the most luminous stars, most of which turned out to be the Hubble-Sandage variables (Chap. 10). By the 1980s instrumentation was adequate for obtaining good spectra of normal high-luminosity stars. Among the first spectroscopic surveys was that of Conti and Massey (1981), who obtained spectra of Wolf-Rayet stars in M33. Many were found to lie within the confines of luminous HII regions, such as NGC 604. The superluminous WR stars resemble similarly-luminous WR stars in the LMC, while the others, which are found within small HII regions, are similar to the common WR types found in the MWG. Further discoveries were reported in Massey and Conti (1983), in which it was reported that the ratio of WC to WN type WR stars decreases with distance from the center of M33, suggesting a gradient in either the abundances or the initial mass function of massive stars. Additional WR stars were reported by Massey et al. (1987).

P. Hodge, *The Spiral Galaxy M33*, Astrophysics and Space Science Library 379,
DOI 10.1007/978-94-007-2025-1_8,

Spectra of medium resolution for early-type M33 stars were obtained in the 1980s by Humphreys et al. (1990). Using the multiple mirror telescope on Mount Hopkins and the 2.1-m telescope on Kitt Peak, the authors determined spectral types of six stars that were found to be single-star members of M33. All were early type stars, ranging in spectral type from B0–B1 II–III to A5e Ia. Photometry of these stars allowed the determination of reddening and total luminosity for each. The range in bolometric magnitude was from −10.02 to −6.2.

Ultraviolet spectra obtained using International Ultraviolet Explorer satellite, together with visible wavelength spectra, were used by Bianchi et al. (1991) to study five hot stars in M33. The stellar winds were found to be weak and the extinction law was found to be similar to that of the Large Magellanic Cloud. Some luminous O and B stars in M33 were observed by Massey et al. (1995) (Fig. 8.1), who compared the sample with that from M31 and NGC 6822. Spectra indicated that these are massive stars with masses greater than ~80 solar masses and luminosities greater than −11 (bolometric).

Full abundance analysis for M33 stars awaited the use of larger telescopes and high resolution spectroscopy. McCarthy et al. (1995) obtained spectra of two A-type supergiants in M33 using the high resolution echelle spectroscope of the 10-m Keck telescope on Mauna Kea, Hawaii. One of the stars, 117-A, was found to have a metal abundance less than solar, while the other, B-324, appeared to have a solar abundance. The fact that the first of these is at a considerably greater distance from the center of M33 than the second supports the fact that the galaxy has an abundance gradient (see Sect. 8.4). B-324 shows P Cygni profiles and analysis of the profiles indicates a stellar wind with a mass-loss rate of 1.2×10^{-5} solar masses per year. The star 117-A has a much less strong wind, with a mass loss rate of 2.5×10^{-8} solar masses per year (Fig. 8.2).

Additional spectroscopy of stars in M33 was carried out in 1994 by Monteverde et al. (1996) using the 4.2-m William Herschel Telescope at La Palma in the Canary Islands. The spectral types of the ten stars ranged from B1 Ia to F0/F5 1a (Fig. 8.3). Luminosities for the stars were determined from the H-alpha profiles, which allowed an independent determination of luminosities and metallicity. There was a range of metallicity among the stars, from solar to values close to 1/10 solar. As in the case of the McCarthy et al. (1995) paper, the results were consistent with the presence of a gradient in which the metallicity decreases outwards from the center of the galaxy.

The same authors further analyzed the spectra obtained with the William Hershel Telescope in order to determine the abundance of oxygen in the stars (Monteverde et al. 1997). The star names, their spectral types, their temperatures, and their oxygen abundances are shown in Table 8.1, in which the results for the two stars observed by McCarthy et al. (1995) are appended. An indication of a gradient with central distance was again indicated.

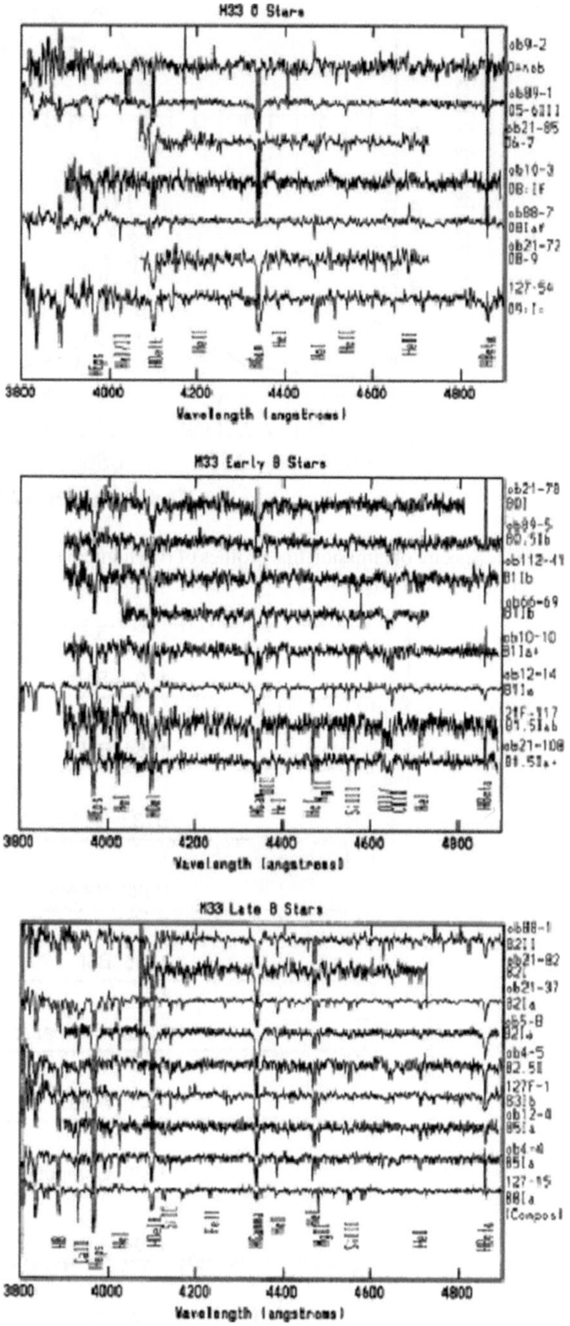

Fig. 8.1 Spectra of luminous blue stars in M33. Names and spectral types are given at the *right margin* (From Massey et al. 1995). Used by permission, copyright AAS

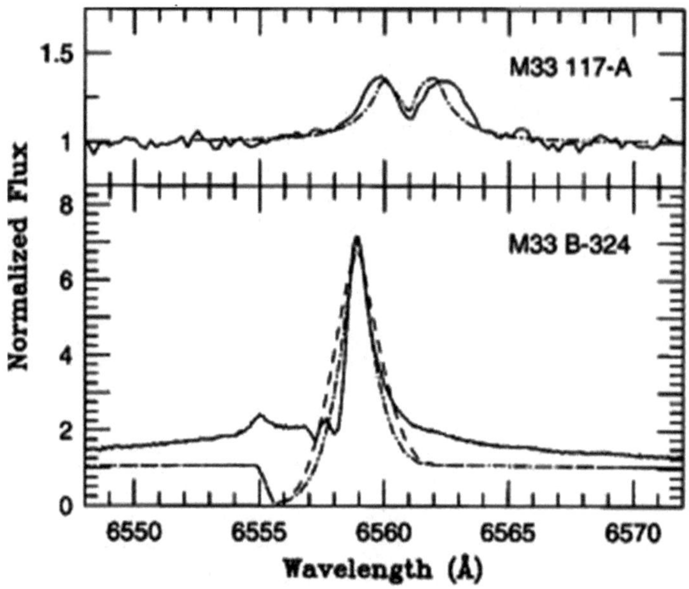

Fig. 8.2 Spectra of emission line profiles of two luminous M33 stars (McCarthy et al. 1995). Used by permission, copyright AAS

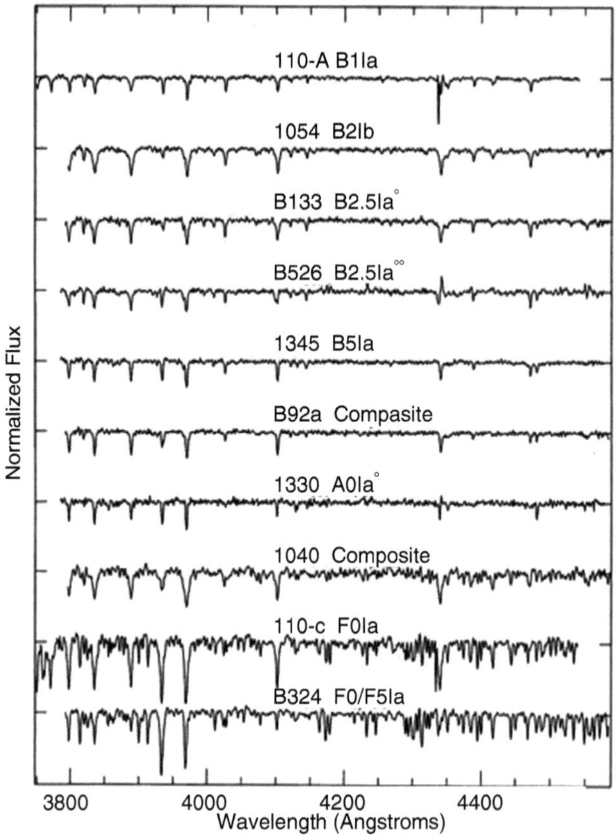

Fig. 8.3 Spectra of stars in M33 obtained by Monteverde et al. (1996). Used by permission, copyright ESO

Table 8.1 Oxygen abundances in B-Type stars in M33 (From Monteverde et al. 1997)

Star name	Spectral type	Effective temperature (K)	Oxygen abundance 12 + log(O/H)
1054	B2 Ib	20,000	9.03
0515	B2.5 Ia	17,500	
1345	B6 Ia	12,500	8.50
0785	B1 Ia	22,500	
B133			8.56
B110A			8.43

8.2 Stellar Photometry

Additional information on the chemical abundances of stars in M33 has been provided by color magnitude-diagrams of individual stars. In the main disk M33 this type of data has primarily come from color-magnitude diagrams of star clusters, as discussed in Chap. 6. Most information available at this time is for the globular star clusters, including both classical globulars and rich clusters of intermediate age (5–7Gyr). As explained in Chap. 6, M33 has a population of low abundance clusters of intermediate age that is different from the population of clusters in the Galaxy. Generally, color-magnitude diagrams of clusters in the disk are found to have abundances midway between that of the Milky Way and that of the Small Magellanic Cloud.

A recent and very thorough investigation of the stellar content of M33, using the Advanced Camera for Surveys on the Hubble Space Telescope, was carried out by Williams et al. (2009). Although they obtained data on the abundances at four different positions along the major axis of the galaxy, their first paper does not give information on the metallicities derived, which is promised for a later paper.

Most multicolor stellar photometry of M33 was carried out in the outer regions because of the smaller crowding of stellar images. A pioneering attempt to determine the abundances in the halo stars in M33 was made by Mould and Kristian (1986). Using the Palomar 5-m telescope and an early CCD detector they obtained a color magnitude diagram of an outer field of M33 to a magnitude faint limit of I = 22.5. The results showed a giant branch that was interpreted to be made up of low metal abundance (Fe/H = –2.2), but with a fairly large uncertainty.

Davidge (2003) used the 8-m Gemini telescope on Mauna Kea to examine stars in the outer regions of M33, finding a somewhat different result. The mean metallicity for the red giant branch stars was found to be –1.0 +/–0.3. Also found was a conspicuous red giant clump of AGB stars, indicating a predominance of intermediate age stars at these positions.

A similar ground-based study of the outer regions was carried out using the WIYN 3.5 m telescope on Kitt Peak by Tiede et al. (2004). Their photometry reached an I magnitude of approximately 25 and clearly showed the red clump of core helium burning stars near the photometry's limit. The magnitude of the red giant clump provided a distance of 867 +/–28 kpc, close to our adopted value of 850 kpc. Using Milky Way globular clusters for comparison, the authors derived a mean metallicity of [Fe/H] –1.0, larger than the metallicity found in the Mould and

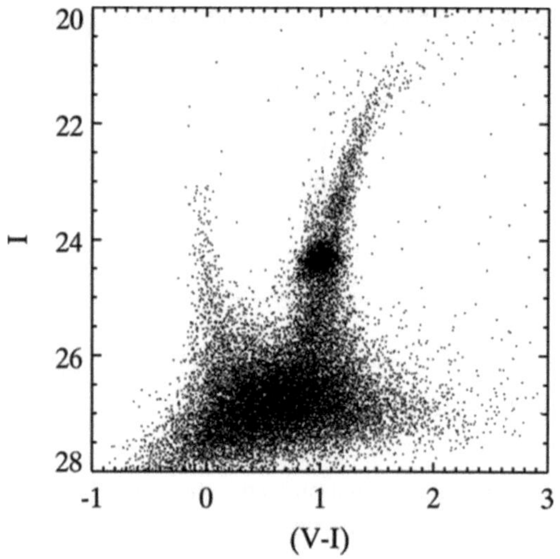

Fig. 8.4 The color-magnitude diagram of the innermost field of M33 studied by Barker et al. (2007), showing the main sequence, the *red giant branch* and the *red clump*. This diagram was found to indicate a mean metellicity of [Fe/H] of −1.14. Used by permission, copyright AAS

Kristian paper, but in agreement with the values derived for disk stars closer to the nucleus. They concluded that even at a distance of 10 kpc from the center of M33 the majority of the stars are disk stars.

Using the Advanced Camera for Surveys on the Hubble Space Telescope, Barker et al. (2007) reached farther into the outer regions of M33, where they obtained V, I photometry. The field examined included the smaller field observed by Tiede et al. (2004) and had a mean distance from the center, deprojected, of 11 kpc. The color magnitude diagram revealed a mix of ages from approximately 100 Myr to several Gyrs. Figure 8.4 is a color magnitude diagram for the innermost field in which a blue main sequence, a red giant branch and the red clump are all clearly shown. The outermost of their fields is shown in Fig. 8.5 in which the younger stars are very few, although the red clump and giant branch are still conspicuous. An elaborate and thorough analysis, using a grid of Milky Way globular clusters of known abundances, provided a metallicity distribution function for each field. The mean metallicities were found to be −1.14 for the inner field, −1.23 for the intermediate field and −1.30 for the outer field, with uncertainties of approximately 0.5 dex. The conclusion of this study is that the outer regions of M33 have about 1/10 the solar metallicity and that there is a radial gradient outwards, as discussed in Sect. 8.4.

8.3 Abundances from Emission Regions

Although of use only for a limited number of atomic species, spectrophotometry of emission regions can provide measurements of abundances in extragalactic HII regions and planetary nebulae. Aller (1942) obtained the first spectroscopy of H II regions in M33, obtaining measurements of the oxygen abundances for several HII regions. Another early survey of emission spectra of giant HII regions in

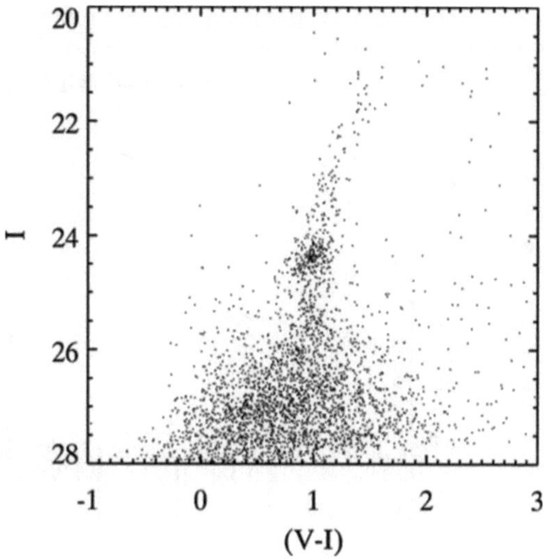

Fig. 8.5 The color-magnitude diagram of a more remote outer region of M33, found to have a metallicity of −1.30 (Barker et al. 2007). Used by permission, copyright AAS

Table 8.2 Abundances in two HII regions in M33 (After Vilchez et al. 1988)

Region	NGC604	NGC588
12+log O/H	8.02	8.30
12+log N/H	7.35	6.77
12+log S/H	6.95	6.86
He/H×100	8.3	8.3

galaxies was carried out by Smith (1975) at the Lick Observatory. Included were 12 HII regions in M33. Line ratios were obtained for hydrogen, carbon, nitrogen, neon and helium. Smith was able to detect and measure the [OIII] line at 4363 Å, which allowed the determination of the electron temperatures and thus the true abundance of oxygen and the other elements. Smith's primary goal was to obtain information on the gradient of abundance in late type galaxies (Sect. 8.4).

A more recent study of the giant HII regions in M33 was carried out by Vilchez et al. (1988). The selected emission lines included [OII], [OIII], [SII], [SII], [NII], HeI and Hα. The abundances of those elements are given for two examples of HII regions in Table 8.2.

The neon abundance in M33 HII regions is of interest because neon is believed to be formed in stars just prior to a supernova phase. A survey of 25 of M33's HII regions was carried out using the Infrared Space Observatory (Willner and Nelson-Patel 2002). They found that the neon abundance in the HII regions ranged from log(Ne/H) +12 = 7.7 to 7.9. The O and Ne abundances in M33 emisson regions were further examined by Crockett et al. (2006). The Kitt Peak Observatory 4 m telescope provided observations that included the [OIII]4363 Å line so that electron temperatures could be obtained. Included in their sample were two HII regions in which the excitation temperature was sufficiently high that ionized helium lines were detected (Fig. 8.6). Table 8.3 summarizes their results for the O and Ne abundances of the centrally-located nebulae.

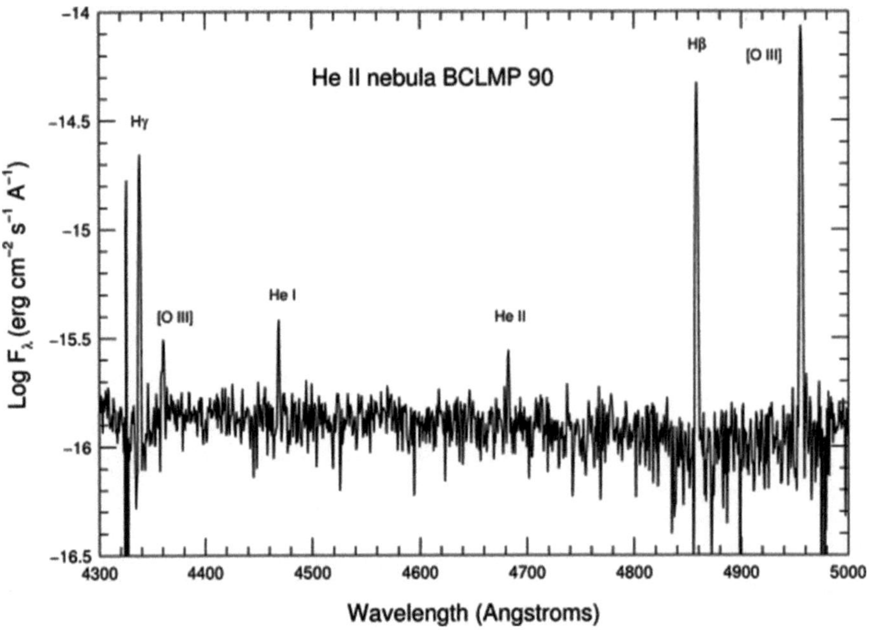

Fig. 8.6 A section of the spectrum of Boulesteix HII Region 90, obtained by Crockett et al. (2006), using the KPNO 4 m telescope. The *helium lines* are indicated. Used by permission, copyright AAS

Table 8.3 Abundances of O and Ne in HII regions (After Crockett et al. 2006)

Name (Boulesteix et al.)	12+log O/H	12+log Ne/H
90	8.50	7.96
691	8.26	7.62
745	8.07	7.2
706	8.32	7.6
290A	8.21	7.4

The results of many studies of the abundances in HII regions for M33 are in very reasonable agreement for the brightest HII regions. The oxygen abundance averages (12+log O/H)=~8.4 for the central regions, with a spread of about 0.11, and varies shallowly with distance from the center (see Sect. 8.4). The neon abundance correlates closely with the oxygen abundance, and the other alpha elements show a similar pattern of abundance ratios to those in the Milky Way HII regions.

Although planetary nebulae are astrophysically different from HII regions, recent spectrophotometry of M33's planetary nebulae shows abundances and abundance gradients that are generally in agreement with those for HII regions (Magrini et al. 2009, Bresolin et al. 2010). Because the majority of the planetary nebulae are evolved from older stars (over 0.3 Gyr in age), the tight correlation between oxygen and neon abundances indicates that oxygen has not been enriched in the interval since the progenitor stars were formed. The abundance of argon also correlates with the oxygen abundance. These results indicate that these elements have not had their surface abundances affected by dredge up from the interiors of the stars during the asymptotic giant branch phase.

8.4 The Abundance Gradient

The chemical abundances of stars and interstellar gas in spiral galaxies have been found to decrease outwards. This effect was first noticed by Kwitter and Aller (1981), who measured the oxygen abundance and found a gradient of –0.13 dex/kpc. Vilchez et al. (1988) confirmed the effect and found that it was steeper in the inner portion of M33 than in its outer regions, with an overall average of –0.12 and a gradient of –0.06 for the more remote nebulae. The Infrared Space Observatory was used in 2002 to measure neon abundances in M33 (Willner and Nelson-Patel 2002) with a very different result. The gradient was found to be –0.15 in the inner regions but steeper in the outer regions with a value of –0.35. In another survey Magrini et al. (2007) also found a difference in the slope, with a much smaller overall gradient –0.054, and with a central value of –0.19 and an outer flatter slope of –0.038. Also recently, Bresolin et al. (2010) obtained an overall gradient of –0.025.

Rubin et al. (2008) used the Spitzer Space Telescope to measure the strength of five emission lines ([SIV] 10.51, H(7-6) 12.37, [NeII] 12.81, [NeIII] 15.56 and [SIII] 18.71) from HII regions in M33. Among the many things learned about the HII regions, they measured the gradient of the abundance of Ne, finding a value of –0.058 (+/–0.014) dex/kpc (Fig. 8.7), a value that is somewhat higher than most recent values measured for oxygen.

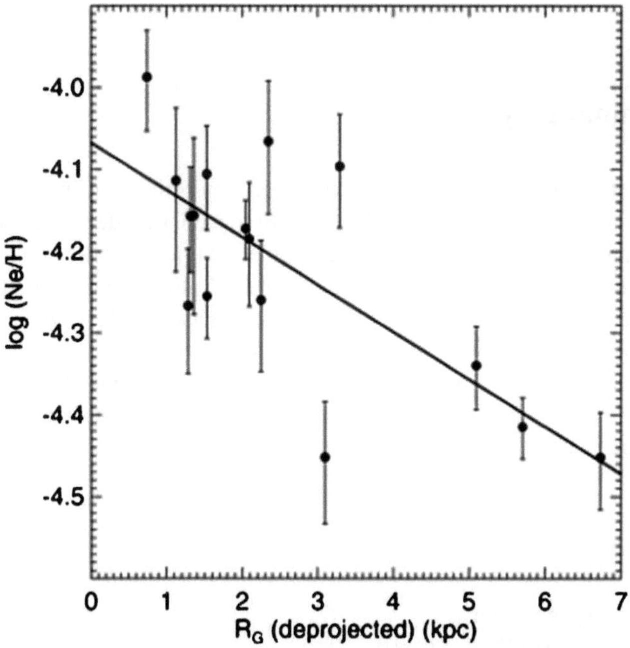

Fig. 8.7 The abundance of neon relative to hydrogen in M33 determined by Rubin et al. (2008) using the Spitzer Space Telescope. Used by permission, copyright AAS

Table 8.4 The oxygen abundance gradient in M33

Source	Location	dex/kpc (deproj) O/H	err
Kwitter and Aller (1981)		–0.13	
Vilchez et al. (1988)	Average	–0.12	
Vilchez et al. (1988)	Outer	–0.06	0.01
Willner and Nelson-Patel (2002)	Inner	–0.15	0.07
Willner and Nelson-Patel (2002)	Outer	–0.35	0.07
Crockett et al. (2006)		–0.012	0.011
Magrini et al. (2007)		–0.031	0.013
Magrini et al. (2007)	Central	–0.19	
Magrini et al. (2007)	Outer	–0.038	
Magrini et al. (2007)	Overall	–0.054	
Rosolowsky and Simon (2008)		–0.027	0.012
Bresolin et al. (2010)		–0.025	0.006
Viironen et al. (2007)		–0.05	0.01

Table 8.4 gathers some of the values obtained for the gradient in the abundance of Oxygen. These results show a large dispersion, most of which is probably the result of the difficulty of the observations and the different way in which the characteristics of the lines are converted to abundance. The data seem to indicate a small slope on the order of –0.1, but with a large uncertainty. The unweighted mean value of the data in Table 8.4 is –0.095 with a standard deviation of 0.095. Of course, a mean of unevaluated data should not be taken too seriously.

8.5 Summing Up

The elemental abundances in M33 have been determined from spectroscopy of bright stars, from color-magnitude diagrams fit to stellar models and from emission line strengths in HII regions and related nebulae. Although there is a spread in the derived or implied values of the metallicity of M33, the general conclusion is that the disk has an [Fe/H] of about −1.0, placing it between the values for the solar neighborhood and the Magellanic Clouds. Many determinations of an abundance gradient with distance from the center of M33 have been published and they show a wide scatter. However, the conclusion seems clear that there is such a gradient and that its value is about –0.1 dex/kpc.

References

Aller, L.H.: ApJ **95**, 52 (1942)
Barker, M.K., et al.: AJ **133**, 1138 (2007)
Bianchi, L., Hutchings, J., Massey, P.: A&A **249**, 14 (1991)

Bresolin, F., et al.: MNRAS **404**, 1679 (2010)
Conti, P., Massey, P.: ApJ **249**, 471 (1981)
Crockett, N.R., et al.: ApJ **637**, 741 (2006)
Davidge, T.J.: AJ **125**, 3046 (2003)
Humphreys, R., Massey, P., Freedman, W.: AJ **99**, 84 (1990)
Kwitter, K., Aller, L.H.: MNRAS **195**, 939 (1981)
Magrini, L., et al.: A&A **470**, 865 (2007)
Magrini, L., Strangellini, L., Villaver, E.: ApJ **696**, 729 (2009)
Massey, P., Conti, P.: ApJ **273**, 576 (1983)
Massey, P., et al.: PASP **99**, 816 (1987)
Massey, P., et al.: AJ **110**, 2715 (1995)
McCarthy, J.K., et al.: ApJ **455**, 135 (1995)
Monteverde, M.I., et al.: A&A **312**, 24 (1996)
Monteverde, M.I., et al.: ApJ **474**, 107 (1997)
Mould, J., Kristian, J.: ApJ **305**, 591 (1986)
Rosolowsky, E. and Simon, J.: ApJ **675**, 1213 (2008)
Rubin, R.H., et al.: MNRAS **387**, 45 (2008)
Smith, H.E.: ApJ **199**, 591 (1975)
Tiede, G.P., Sarajedini, A., Barker, M.K.: AJ **128**, 224 (2004)
Viironen, K. et al.: MNRAS **381**, 1719 (2007)
Vilchez, J.M., et al.: MNRAS **235**, 633 (1988)
Williams, B., et al.: ApJ **695**, L15 (2009)
Willner, S.P., Nelson-Patel, K.: ApJ **568**, 679 (2002)

Chapter 9
Stellar Populations and the Star Formation History

9.1 Recognizing the Two Populations

An early determination of the stellar populations in M33 was the analysis of the stellar photometry obtained by Mould and Kristian (1986), as described in Chap. 8. Their results indicated that the galaxy contains a population of old stars, similar to those in the halo of our Galaxy. A second indication of the presence of a very old, population II component was the discovery of RR Lyrae stars (Pritchet 1988), as described in Chap. 10. A population of old globular clusters, studied both by color magnitude diagrams and kinematics (Schommer et al. 1991), further indicated the presence of an old population (Chap. 6).

A more extensive photometric study of the stellar populations of M33 was carried out by Wilson and Scoville (1990), who used the Canada – France – Hawaii Telescope to study four fields of the disk of M33. The photometry extended to a limit of V=21 and it clearly showed both a young main sequence of blue stars (Population I) and a field of red giants (Fig. 9.1). The authors identified, in addition to the young stars on the MS and intermediate age supergiants, an old giant population from which they calculated a distance to the galaxy from the red giant tip.

9.2 The Central Area of the Disk

Images from the Hubble Space Telescope were used by Mighell and Rich (1995) to measure the stellar populations of the central area of M33. Using visual and near infrared passbands, they displayed a color-magnitude diagram with a conspicuous main sequence, an old red giant branch, an area of red supergiants and some blue loop stars (Fig. 9.2).

From the wide red giant sequence it was possible to conclude that there is a considerable abundance spread among the red giants, on the order of a factor of 30. The red giants were found to be centrally concentrated, in contrast to the distribution

P. Hodge, *The Spiral Galaxy M33*, Astrophysics and Space Science Library 379,
DOI 10.1007/978-94-007-2025-1_9,

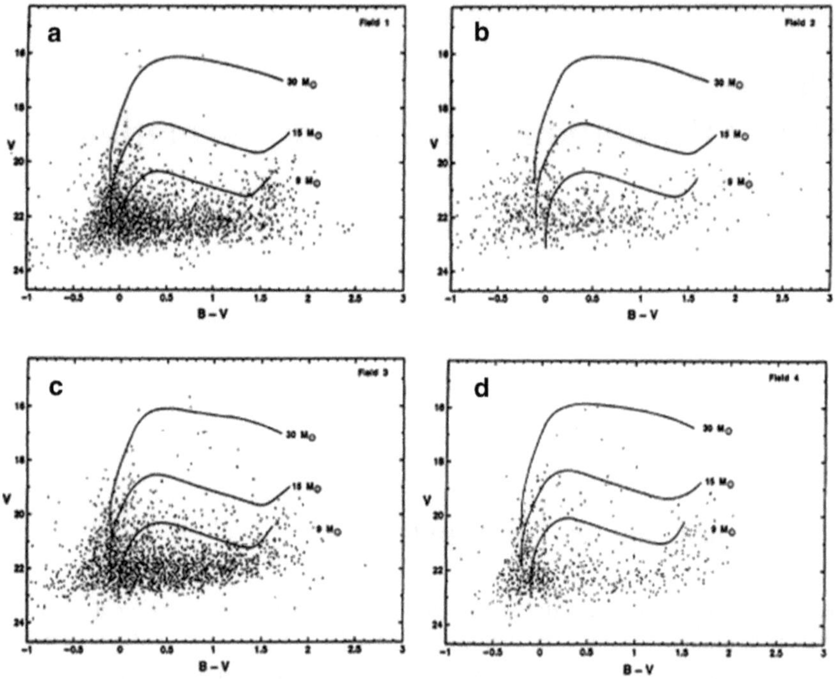

Fig. 9.1 Color-magnitude diagrams of four M33 disk star fields (From Wilson and Scoville 1990). Used by permission, copyright AAS

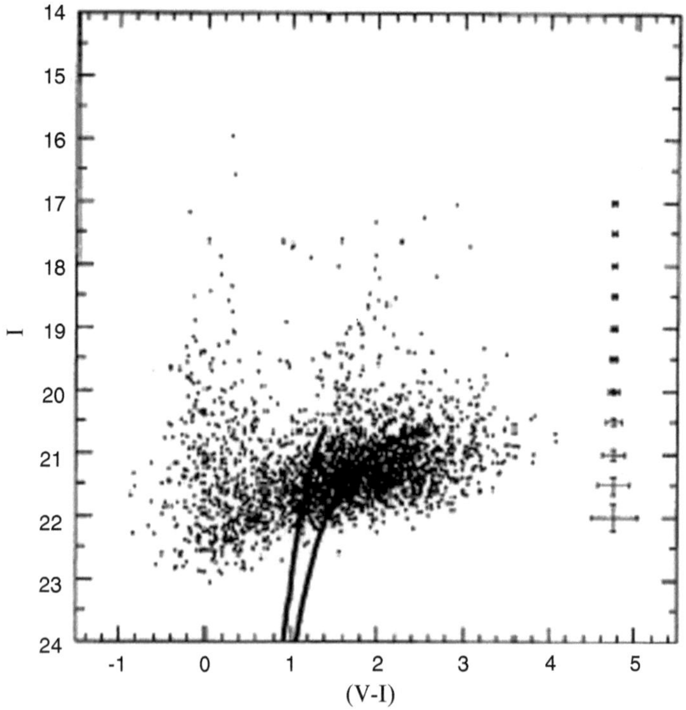

Fig. 9.2 A CMD for M33 from a HST study by Mighell and Rich (1995). Model giant branches are shown as *thick lines* and uncertainties are shown on the *right*. Used by permission, copyright AAS

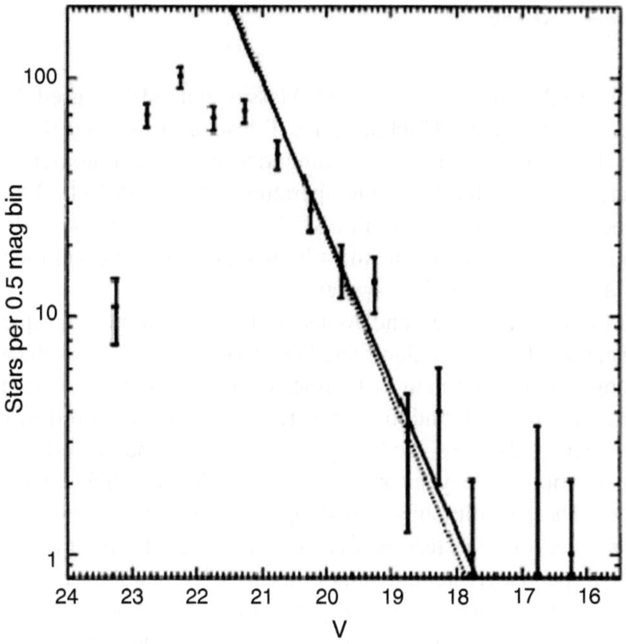

Fig. 9.3 The luminosity function of stars in the disk of M33 compared to two power law fits (From Mighell and Rich 1995). Used by permission, copyright AAS

of younger stars. These stars were suggested to be part of the central bulge of M33, a subject of controversy (Minniti et al. 1993). The blue stars were identified as normal main sequence stars and the high luminosity of the top of the sequence indicated a young age component. The luminosity function of the main sequence is shown in Fig. 9.3. The basic star formation history proposed on the basis of the color magnitude diagram consists of three periods of star formation: an initial period in which the putative bulge formed a population of old metal poor stars, a later period, during which the current red clump stars and yellow supergiants were formed, and a recent period in which the young stellar associations were formed.

A different approach to the problem of stellar populations in the central part of M33 was used by McLean and Liu (1996), who obtained infrared photometry of the galaxy. The results were color-magnitude diagrams and luminosity functions in the infrared JHK system. A conspicuous component of the diagrams was a population of luminous giants that were found both in the core region and in the surrounding annulus of disk stars. The authors concluded that the stars were intermediate age supergiant stars belonging to the disk population and that there was no evidence of a bulge of older stars in the central core.

9.3 Massive Stars

Concentrating on the young stars in M33, Massey et al. (1996) used the Ultraviolet Imaging Telescope and the Hubble Space Telescope to survey the stars of high temperature (Fig. 9.4). The survey was limited to stars with bolometric magnitudes brighter than −9, and with effective temperatures of 10,000–15,000 K. These stars neatly outline the spiral arm pattern of M33 and extend over the entire disk, right into the central core. Several of the most luminous stars in the sample were proven by the HST images to be multiple systems.

Massey (1998) and Massey and Johnson (1998) examined the question of the number and nature of red supergiants and WR stars in M33. The results were instructive in the ongoing development of theories of the late stage evolution of massive stars. In particular, it was found, by comparing the numbers of red supergiants and WR stars in the three galaxies, the Milky Way, M33, and the Magellanic Clouds, the relative numbers of the two types of stars were correlated with the metal abundances in the galaxies. These results suggest that higher metallicities result in a higher mass loss rate for the evolving stars so that a higher metallicity star of a particular luminosity will spend a greater amount of time as a WR star than as a red supergiant. Further spectroscopic work on the WR population was carried out by Abbott et al. (2004), who refined some of the classifications reported on previously.

Spectroscopy of several massive stars in M33 was carried out by Bianchi et al. (2001). Stars were selected by wide-band photometry using HST images. Fitting by

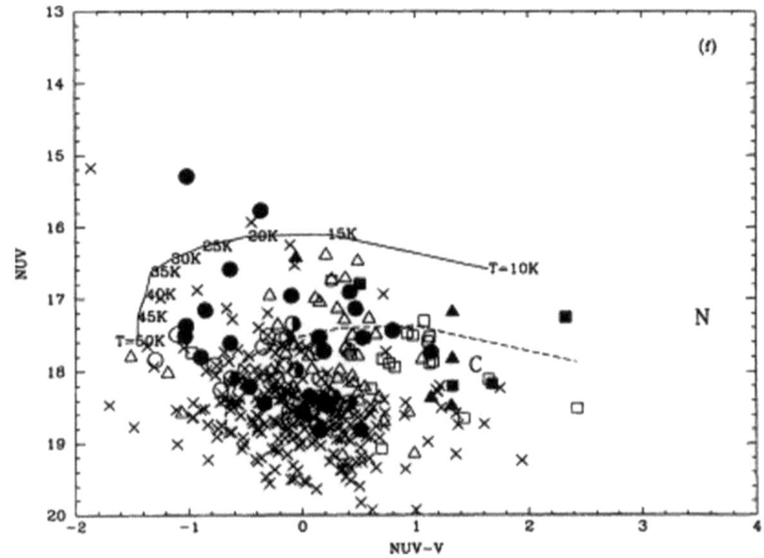

Fig. 9.4 The M33 CMD including ultraviolet magnitudes, showing the hottest stars. Temperature are indicated along the most massive evolutionary track. *Symbols* represent different spectral types and N is the M33 nucleus (From Massey et al. 1996). Used by permission, copyright AAS

a combination of Balmer line intensities and the wide-band photometry provided values for effective temperature and gravity. Temperatures of the stars ranged from approximately 18,000–30,000 K.

A study of carbon stars was carried out by Block et al. (2004), who used the Palomar Mountain 5-m telescope to obtain near-infrared photometry of M33. They discovered large numbers of carbon stars in the outer regions, confirming the presence of intermediate age stars in the outer disk.

A complementary study of carbon stars in M33 was made by Rowe et al. (2005), using the Canada–France–Hawaii telescope on Mauna Kea. Using both broadband photometry and narrowband imaging, it was possible to separate the AGB population into carbon stars and M giants. The carbon stars extended into the outer regions, agreeing with previous detections; the authors interpret the carbon star distribution as being simply a continuation of the disk of the galaxy.

9.4 The Outer Disk and Halo

Many important papers published in the first decade of the twenty-first century described the determination of detailed patterns of star formation rates in M33. Because of photometric difficulties in the crowded central regions, several of these concentrated on the outskirts, where, in any case, important questions existed regarding the presence or absence of a halo population. The first of these was a survey that used the Gemini North Telescope on Mauna Kea. Davidge (2003) obtained four-color data for a field covering a line-of-sight distance from 8 to 10 kpc. He found both a normal old red giant branch and a similarly prominent population of AGB stars. The AGB stars are part of an outer intermediate-age population, which is far outside of the main disk of M33. There are also intermediate-age star clusters at these distances and it is likely that the two have the same history. Finding such a population in what for our Galaxy is a pure old halo indicates that M33 formed differently, probably more gradually, especially in its outer parts.

A paper that confirmed and extended the Davidge result was presented by Galleti et al. (2004), who obtained V and I photometry of two outer fields in M33, covering distances from 10 arcmin to 33 arcmin (2.5–8 kpc) from the center. In their fields the AGB was found to show a peak at Fe/H = ~−1.0, while the red giant branch magnitudes indicated metallicities down to M/H = −2.0.

Much of our recent knowledge of M33's star formation history has resulted from the research of a group of astronomers led by Ata Sarajedini at the University of Florida. In addition to the series of papers on M33's clusters mentioned in Chap. 6, there is a series of papers on the stellar populations in M33's outer regions, as shown in Fig. 9.5 (Tiede et al. 2004; Barker et al. 2007a, 2007b; Barker and Sarajedini 2008). These papers have also been discussed in Chap. 8, Sect. 8.2. Their color-magnitude diagrams indicate that there was an extensive star-forming period approximately 10 Gyr ago, which explains the low metallicity (M/H = −2), old red giant branch.

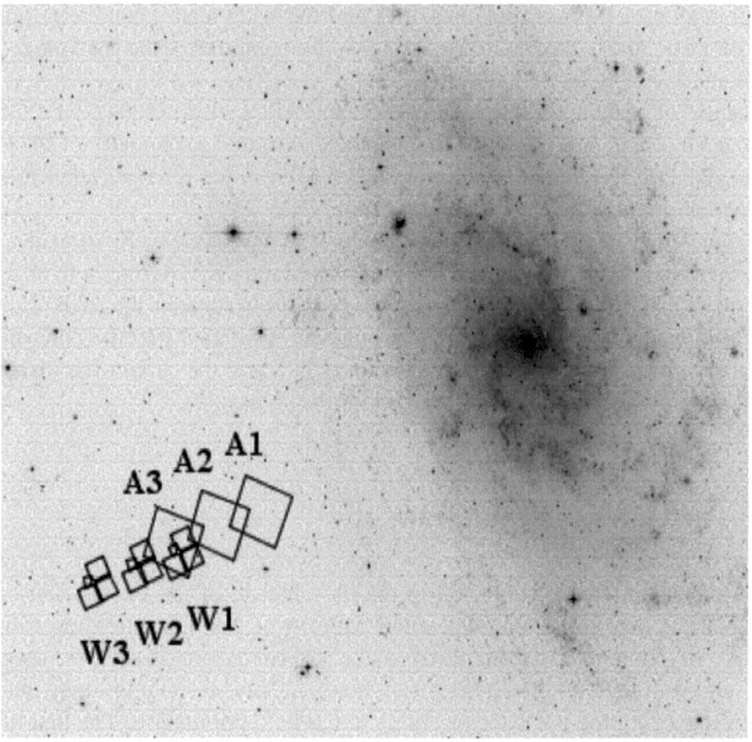

Fig. 9.5 Locations of the outer regions of M33 observed by the University of Florida group (Barker 2007a). Used by permission, copyright AAS

Subsequent star formation activity produced stars with enriched (M/H = −1) composition. Both components extend over at least 10 kpc (deprojected) from the center.

Figure 9.6 shows an example of the Barker et al. (2007b) results for the star formation history of the inner of their fields (~20 arcmin from the center). In this figure, panel (a) shows the derived star formation history, (b) shows the cumulative fraction of stars by age, (c) shows the metallicity as a function of age, (d) gives the cumulative distribution of metallicitiy, (e) and (f) display the observed and model CMDs, (g) shows the difference between observed and model CMDs and (h) is a plot of significance. These data were analyzed by using the software developed by Dolphin (2002), which finds a best-fit solution to the star-formation history of a CMD, giving both the rate of star formation and the stellar abundances as a function of time over the age of the population. The program usually assumes a Salpeter (1955) initial mass function (IMF), the Padua stellar evolution models and solves for a mean reddening value. This kind of exhaustive analysis of a galaxy's stellar population had by 2007 become a standard for the field.

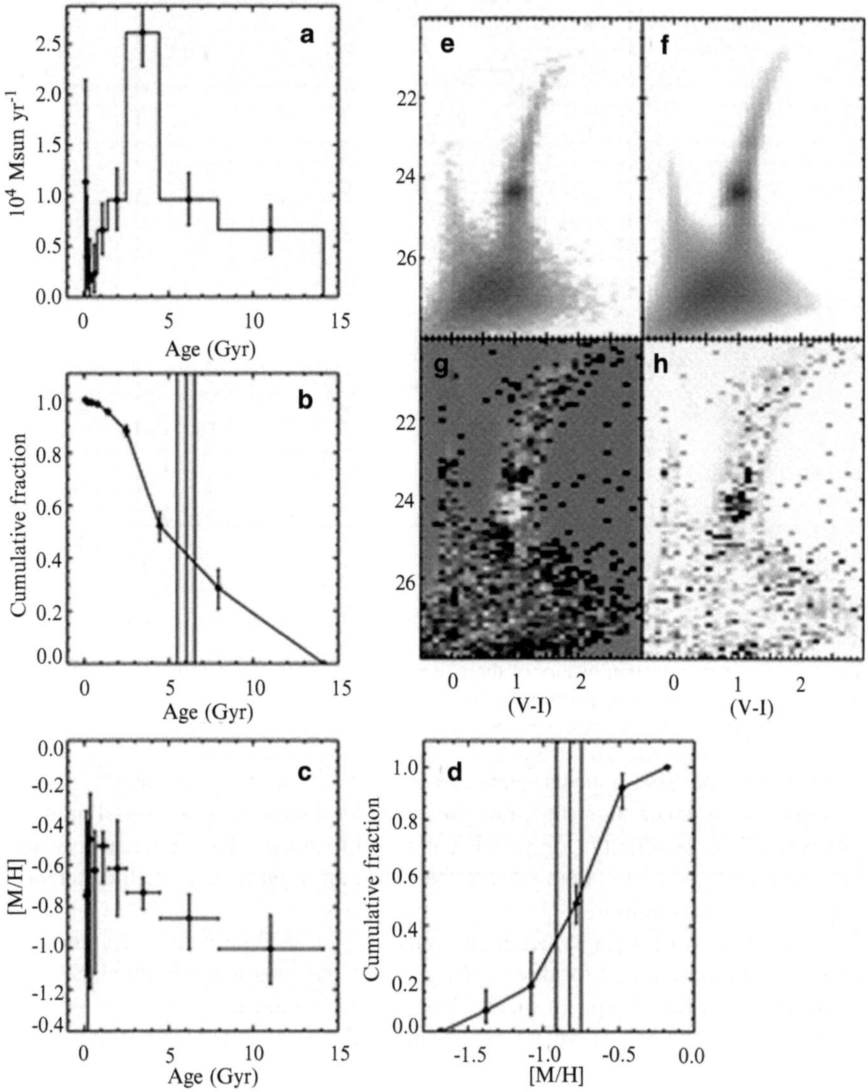

Fig. 9.6 A complete star formation history story in one figure; this is for Area A1 in the outer regions of M33 (see Fig. 9.5), as analyzed by Barker et al. (2007b). (**a**) is the rate of star formation as a function of age, (**b**) is the cumulative fraction of stars formed over the age of the galaxy, (**c**) is the metallicity as a function of age, and (**d**) is the history of the metallicity enrichment. Images (**e**) through (**h**) show the observed CMD, the model CMD, the difference between the two and its significance, respectively. Used by permission, copyright AAS

The Barker et al. (2007b) observations permitted a study of the changes in the population of stars as one moved from 20 arcmin out to 30 arcmin from the center. The differences are displayed in Fig. 9.7. The inner site (field A1) is plotted by a solid line, the middle field by a dotted line and the outer by a dashed line. In the

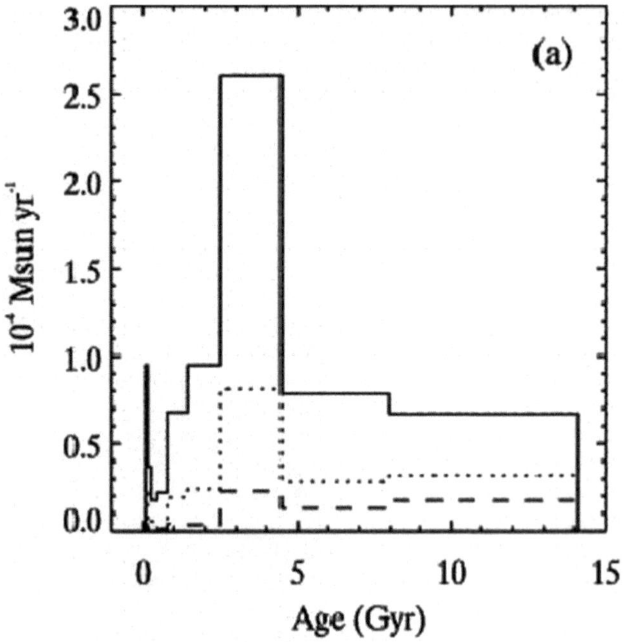

Fig. 9.7 The star formation history of three outer areas of M33 (see text for locations) (From Barker et al. 2007b). Used by permission, copyright AAS

inner region the SFR appears enhanced for ages ~2 to 5 Gyr, with an apparent decrease for younger stars (this, however, may be distorted by the small numbers involved). The metallicity (log [M/H] compared to solar values) decreases from a present value of −0.7 to about −1.0 at the oldest ages, with only small differences among the three positions.

A comparison of the age-metallicity relations for M33 derived by Barker et al. (2007b) with those found for three other galaxies, the solar neighborhood (SN), the SMC and the LMC is shown in Fig. 9.8. The M33 data lie close to or possibly slightly below the LMC curve, not surprising for the outer parts of the galaxy, which have lower abundances than average.

Using their deep color-magnitude diagram obtained with the ACS camera of the HST, Barker and Sarajedini (2008) set out to model the distribution of stars in the color-magnitude diagram under various assumptions about the star-formation mechanism and history. The metallicities in a star-forming region increase because of instantaneous enrichment from supernovae of Types II and Ia. The alpha elements (O, Ne, Mg, Si, S, CA, Ti) are mainly produced in the rapid burning and explosion of Type II supernovae, while most of the Fe is produced in the rarer Type Ia supernovae, which are explosions of binary white dwarf stars, which provides a slower process of enrichment. It was possible to construct synthetic CMDs based on various assumptions, including net gas outflow, gas inflow or closed box (wherein the gas is conserved during the process). The best fits to the observations are for

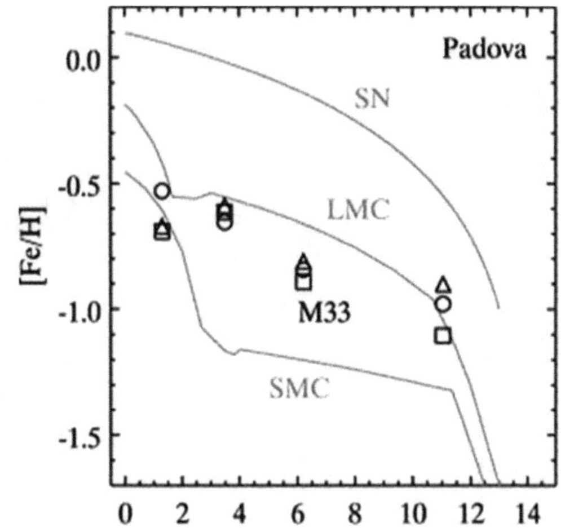

Fig. 9.8 Histories of the metallicities in four galaxies. The *symbols* are M33 results from Barker et al. (2007b), while the *lines* are for the SMC, the LMC, and the solar neighborhood (based on the Padova stellar evolution models). Used by permission, copyright AAS

models in which gas inflow from the surroundings is greater than gas outflow from supernovae. The implications of the results are that M33 formed its disk, including the outer disk, by inflow of gas over several Gyr.

To pursue the question of M33's SFH further, Williams et al. (2009) combined the Barker et al. (2007a, b) images, which were located close to the minor axis, with 4 new HST ACS images taken along the major axis. All seven fields were taken with filters equivalent to V and I of the standard UBVRI system and all were long exposures; the Williams exposures totaled about 6,000 s in the inner location and about 24,000 s for the outer locations. The result is a set of data extending to very faint limits for a large number of stars. A total of 1,537,647 stars had photometry in both colors, producing rich, well-populated color-magnitude diagrams.

The resulting star formation histories of the four Williams fields are shown in Fig. 9.9. A comparison of these curves shows how the star formation rate differs as a function of distance from the center of M33. Figure 9.10 plots the time-dependent cumulative fraction of stars formed for each field. As found in previous studies, the result shows clearly that the inner disk formed most of its stars at an early epoch, about 10 Gyr ago, while the outer disk includes a large fraction of stars formed more recently. This trend is called an "inside–out" disk growth. The percentage mass in stars formed at the early epoch in the innermost field is 71 +/–9%, while this figure decreases outwards to 16+/–6% for the outermost field. The scale length of the disk is found to be a function of age; for the oldest stars, it is 1.0 (+/–0.1) kpc, while for the intermediate age population the scale length is given as 1.8 (also +/–0.1) kpc. There is a tendency for the gradient at very large distances (>~10 kpc) to reverse.

Additional evidence for an inside–out pattern of star formation activity comes from a mutiwavelength study of stars and emission from gas and dust (Verley et al. 2008), who find that the current star formation rate declines outwardly with a longer scale length (~2 kpc) than that for the older population of stars.

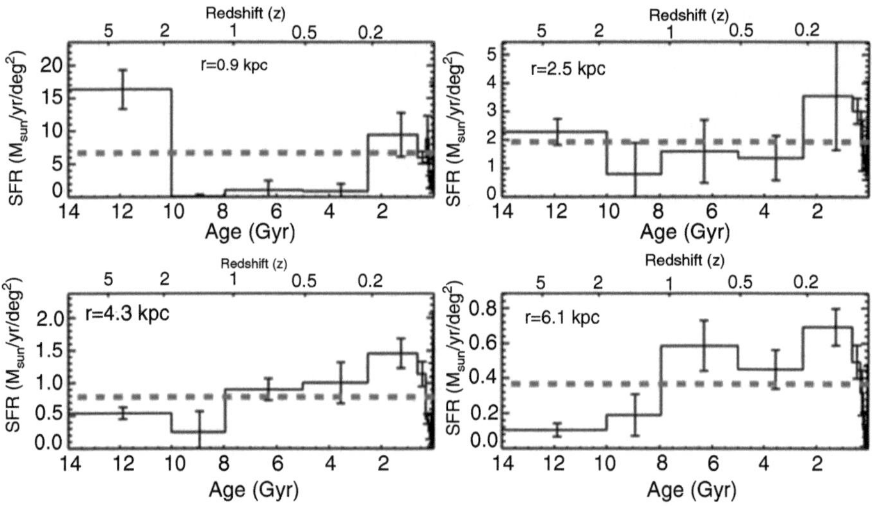

Fig. 9.9 The star formation histories for four positions at various distances from the center of M33 (From Williams et al. 2009). Used by permission, copyright AAS

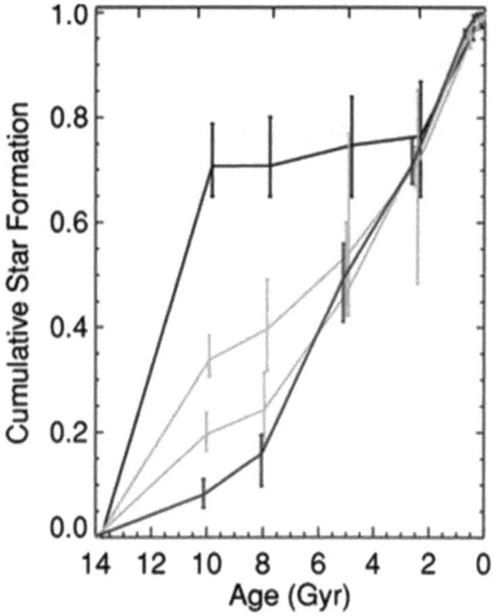

Fig. 9.10 The cumulative star formation in the same four regions as in Fig. 9.9. The *innermost* is the upper curve and the *outermost* is the bottom curve (From Williams et al. 2009). Used by permission, copyright AAS

9.5 The Extreme Outer Population

A survey of the outer regions around M33 discovered a population of extremely distant stellar structures (McConnachie et al. 2010), described in Chap. 4. The stars identified in these remote regions were primarily metal-poor red giants with [Fe/H]

of about −1.6, with an interquartile range of about 0.5 dex. The population appears to belong mostly to the outer disk of M33, rather than to a spherical halo, though this question is not yet clearly settled.

9.6 Summing-Up

M33 contains the full range of stars from dwarfs to supergiants and from just-formed to Hubble-time-old. The pattern of star formation is similar to that for the Milky Way Galaxy, but different in interesting ways. Stars have formed in the disk over the complete range of ages, but with an enhanced rate between 5 and 8 Gyr, when the formation rate reached 2500 solar masses per year in the outer parts of the disk. Most stars in the central areas of the disk were formed at an early era, while in the outer disk, star-formation was predominantly more recent, ergo the "inside–out" scenario. The heavy element abundances increased with time at about the same rate as in the Large Magellanic Cloud, more slowly than in the Galaxy, and reaching a smaller current value. The halo contains old metal-poor stars, but also some of intermediate age and composition. In this way M33's halo is more like M31's than our local Galaxy's.

References

Abbott, J., et al.: MNRAS **350**, 552 (2004)
Barker, M.K., Sarajedini, A.: MNRAS **390**, 863 (2008)
Barker, M.K., et al.: AJ **133**, 1125 (2007a)
Barker, M.K., et al.: AJ **133**, 1138 (2007b)
Bianchi, L., et al.: PASP **113**, 697 (2001)
Block, D.L., et al.: A&A **425**, 37 (2004)
Davidge, T.: AJ **125**, 3046 (2003)
Dolphin, A.: MNRAS **332**, 91 (2002)
Galleti, S., Bellazzini, M., Ferraro, F.: A&A **423**, 925 (2004)
Massey, P.: ApJ **501**, 153 (1998)
Massey, P., Johnson, O.: ApJ **505**, 793 (1998)
Massey, P., et al.: ApJ **469**, 629 (1996)
McConnachie, A.W., et al.: ApJ **723**, 1038 (2010)
McLean, I.S., Liu, T.: ApJ **456**, 499 (1996)
Mighell, M., Rich, M.: AJ **110**, 1649 (1995)
Minniti, D., Olszewski, E.W., Rieke, M.: ApJ **410**, 79 (1993)
Mould, J., Kristian, J.: ApJ **305**, 591 (1986)
Pritchet, C.J.: The Extragalactic Distance Scale. ASP, San Francisco (1988). p 59
Rowe, J.F., et al.: AJ **129**, 729 (2005)
Salpeter, E.: ApJ **121**,161 (1955)
Schommer, R.A., et al.: AJ **101**, 873 (1991)
Tiede, G.P., Sarajedini, A., Barker, M.K.: AJ **128**, 224 (2004)
Verley, S., et al.: Formation and Evolution of Galaxy Disks. ASP, San Francisco (2008). p 91
Williams, B., et al.: ApJ **695**, L15 (2009)
Wilson, C.D., Scoville, N.: ApJ **363**, 435 (1990)

Chapter 10
Variable Stars

In this chapter we review the research that has been carried out on the variable stars in M33. Included in separate sections are the geometrical variables (eclipsing stars) and the various kinds of intrinsic variables.

10.1 Early Studies of Variable Stars in M33

The first variable stars in M33 were a few novae, discovered with the large telescopes at the Mt. Wilson and the Lick Observatories (see Sect. 10.7). A systematic study of variables did not occur until Hubble's (1929) thorough observing program, which discovered a number of different types of variable stars and used the Cepheids among them to establish the extragalactic nature of M33 (Chap. 3). Hubble's data were photographic images taken with blue-sensitive emulsions, giving him no information on the color of the stars. His photometry was based on magnitudes determined by comparison with "standard sequences" of stars in Selected Area 45, which had been measured by fairly crude techniques (compared to modern standards) and which was reliable only to magnitude (pg) = 18.5. His measurements of the magnitudes of the Cepheids reached from pg = 18.3 to >19.5, and so the results depended largely on unreliable magnitudes and were therefore uncertain.

Hubble also discovered and studied what became known as Hubble-Sandage variables, but now more commonly are called luminous blue variables. They are discussed in Sect. 10.5.

It was a long time before anyone looked again at the variable stars of M33. Van den Bergh, Herbst and Kowal (1975) made the next survey of M33 to discover variable stars. Using the Palomar Observatory 48-inch (1.2 m) Schmidt Telescope, they discovered 38 new variables, increasing the total number of known variables to 86. Periods could not be obtained because of crowding of the images on the relatively small-scale plates.

Sandage (1983) attempted to correct Hubble's photometry by remeasuring the standard sequences in Selected Area 45 photoelectrically. He found large errors for

P. Hodge, *The Spiral Galaxy M33*, Astrophysics and Space Science Library 379,
DOI 10.1007/978-94-007-2025-1_10,

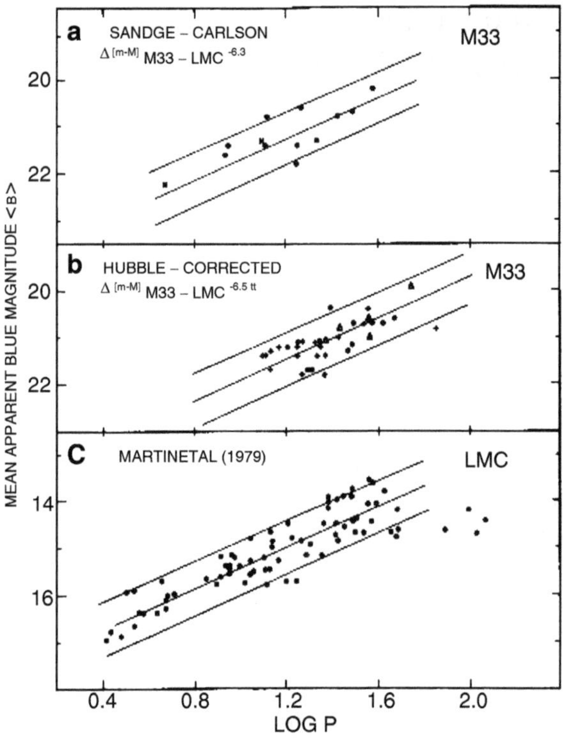

Fig. 10.1 Period-luminosity diagrams for M33 Cepheids. The Sandage-Carlson photometry was the first new data since Hubble's original paper. The P-L relation for the Large Magellanic Cloud is shown for comparison (From Sandage and Carlson 1983). Used by permission, copyright AAS

faint magnitudes, claiming, for example, that the corrections were as large as 2.8 magnitudes at B = 22.5. Additionally, 13 new Cepheids in M33 were discovered and studied by Sandage and Carlson (1983), who produced a new period-magnitude relation (Fig. 10.1).

10.2 Current Large-Scale Searches for Variables

The results of a modern large-scale variable star search were published by Macri et al. (2001). The observations were obtained with CCD cameras mounted on the Michigan-Dartmouth-MIT telescope at Kitt Peak and the F. L. Whipple telescope on Mount Hopkins. A total of 132 nights were devoted to obtaining the images. These data produced nearly 2,000 images in three colors, B, V, and I. The interval of observations was 7.4 years, providing excellent time coverage. The total number of variable stars discovered was 544 and several types of variables were included.

Another large scale survey was carried out using the Mega-camera of the 3.6 m Canada-France-Hawaii telescope on Mauna Kea (Hartman et al. 2006). Using the g', r' i' color system of the Sloan Digital Sky Survey, a total of 36,000 variable sources was detected. The field of view was centered on the nucleus of M33 and had an area of one square degree. The limiting magnitude reached values of approximately 24. The majority of these stars were red and included an estimated 20,000 long period variables (Sect. 10.6) and 2000 Cepheids. These estimates were based on colors, but not confirmed by full analysis of their light curves.

In another survey, using infrared magnitudes, variables luminous in the red were surveyed by using the 3.8-m United Kingdom InfraRed Telescope on Mauna Kea (Cioni et al. 2008). Among the large numbers of AGB stars were 7,650 candidate long period variables.

The Spitzer Space Telescope also was used to make an infrared survey for variable stars, using three wavelengths: 3.6, 4.5 and 8.0 μm (McQuinn et al. 2007). Of the 37,650 stars measured, 2,923 variable stars were identified. Many were expected to be long period variables.

10.3 Eclipsing Binaries

Hundreds of eclipsing binaries have been discovered in M33 in recent years (Macri et al. 2001). Figure 10.2 shows examples of light curves for four of them.

A full analysis of the light curves and orbits of only three eclipsing variables has been published at the time of this writing. The first and most intensely studied is the X-ray eclipsing binary M33 X-7 (Dubus et al. 1999). This binary, remarkable because it is apparently a rare example of a binary that includes a pulsar, is discussed in Chap. 11. The second was discovered in 2009, has a period of 1.732479 days and consists of a neutron star and a hot supergiant. It is also discussed in Chap. 11.

The other well-studied eclipsing binary was one of the 237 candidate binaries found by the DIRECT project (Macri et al. 2001a, b). It has a period of 4.8935 days and contains two luminous and massive stars located in a star-forming region in the OB association No. 66 in the Humphreys and Sandage (1980) catalog. The stars are both of spectroscopic type O7 and they have masses of 33 and 30 solar masses and radii of 12 and 9 solar radii, respectively. Temperatures of the stars are 37,000 and 36,000 K, respectively. The multicolor photometry allows a flux calibration that determines the distance to this binary system of 964 ± 54 kpc, which is about 115 kpc larger than the accepted distance. The discrepancy is not, of course, a suggestion that the stars are far behind M33, but rather that there is a systematic error in one of the two distance determinations.

10.4 Cepheid Variables

The Cepheids are the most thoroughly studied of M33's variables because of their importance in the use of the Cepheid period-luminosity relationship to determine distances to galaxies. That was the reason why Hubble's paper emphasized them and why almost all research on M33's variable stars was devoted to Cepheids. Because precise, reddening-free magnitudes are necessary for determining the P-L

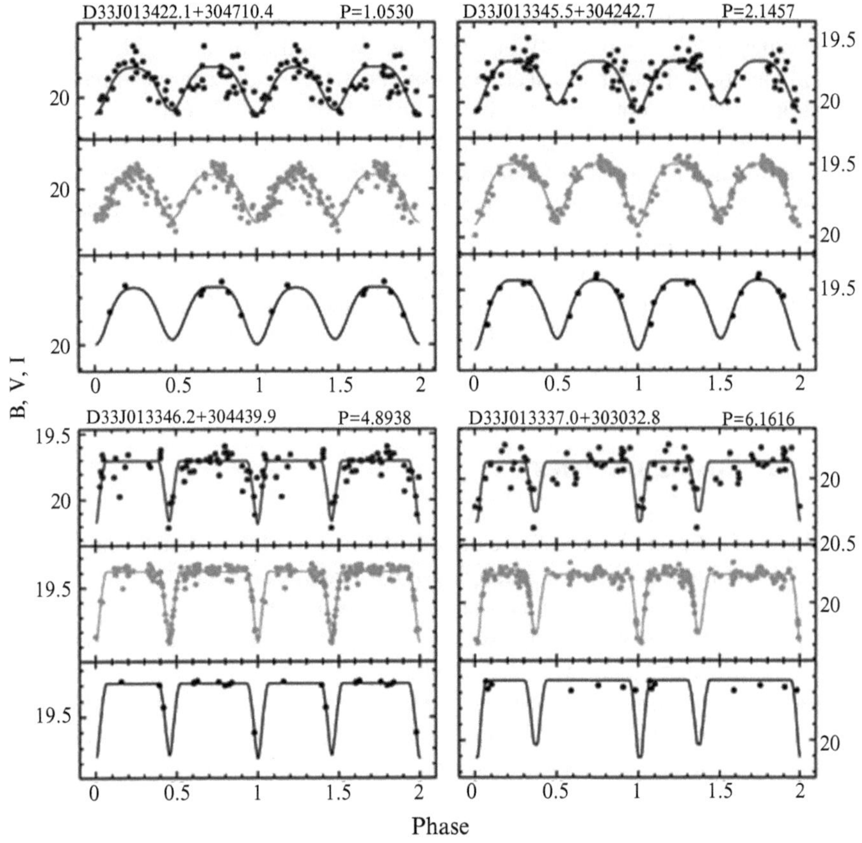

Fig. 10.2 Light curves for four eclipsing binaries in M33. From the *top down*, the curves are for I, V and B filters, respectively (From Macri et al. 2001b). Used by permission, copyright AAS

relation, the amount of extinction for each star must be known, and that requires measurements at a range of colors. In particular, measurements at infrared colors, which are less sensitive to extinction than visible colors, are very useful in obtaining reddening-free magnitudes. The first such magnitudes for M33 Cepheids were obtained by Madore et al. (1985) and they led to an improved value of the distance modulus M33 (see Chap. 1).

The first large-scale search for Cepheid variables beyond those initially discovered by Hubble was made at the Kitt Peak Observatory by Kinman et al. (1987), using photographic plates at the prime focus of the 4-m telescope. Nearly 400 variable stars were identified, 90 of which were Cepheids for which periods could be obtained. Figure 10.3 shows their results for the period-color relation, in which the Cepheids are clearly separated in color from the LPVs.

A new campaign to establish a P-L relationship for M33 was carried out by Freedman et al. (1991), using telescopes at Palomar, Kitt Peak and Mauna Kea.

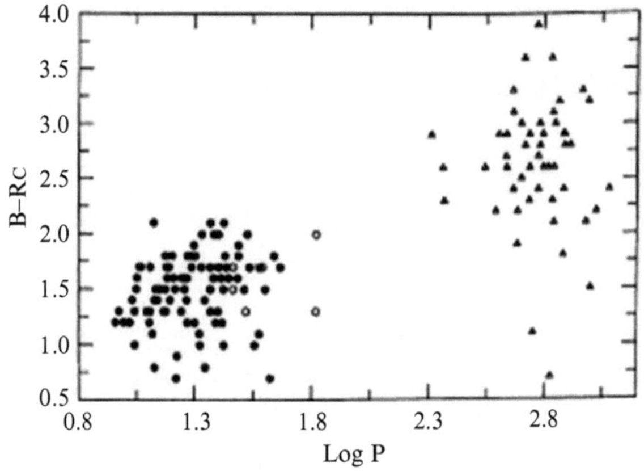

Fig. 10.3 The distribution of colors of three kinds of variables in M33, plotted against log of the period. Cepheids are solid dots, W Virginis variables (a rare type of Population II variable) are *open circles* and long period variables are *triangles* (From Kinman et al. 1987). Used by permission, copyright AAS

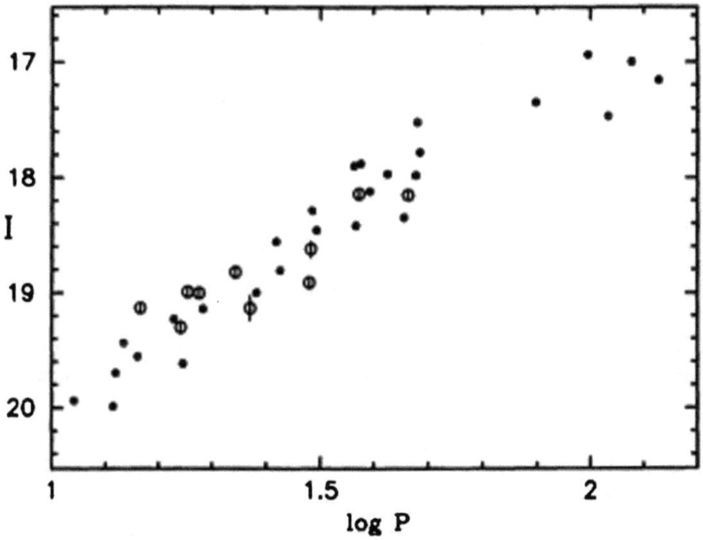

Fig. 10.4 The period-magnitude diagram for Cepheids for I magnitudes. The *open circles* are for Cepheids in M33 and the *filled circles* are for LMC Cepheids, adjusted to fit in magnitude (From Freedman et al. 1991). Used by permission, copyright AAS

Over a 5 year interval the authors obtained a large series of CCD observations, mostly in four colors, using B, V, R, and I filters. A total of 19 Cepheid variables, mostly from Hubble's list, were observed in four different fields, 3 by 5 arcmin in size. The periods for the Cepheids ranged from 3.2 to 46 days and the mean magnitudes were all between B = 20.0 and 21.4. The principal result of the study of the Cepheids was a period-luminosity relation that was determined for each of the four colors (see the example for I magnitudes in Fig. 10.4). The distance modulus

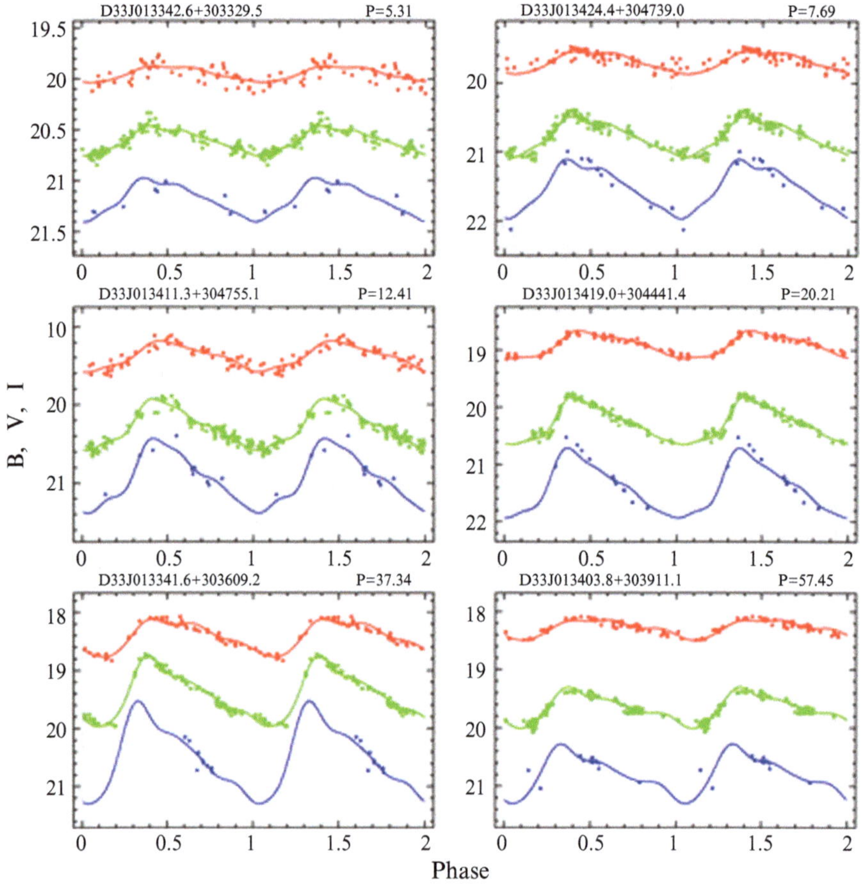

Fig. 10.5 Light curves for 6 Cepheid variables in M33. From the top down, the curves are for I, V and B filters, respectively (From Macri et al. 2001b). Used by permission, copyright AAS

differed for the four different colors as a result of the extinction caused by dust and its corresponding reddening. The mean value of the reddening for the Cepheids was found to be $E(B - V) = 0.10 \pm 0.09$. When corrected for reddening, the linear distance for M33 was determined to be 840 kpc, very close to our adopted current value of 850 kpc.

A deep survey of the central area of M33 was the Macri et al. (2001) photometry described in Sect. 10.2. Of the 544 variables discovered, 251 were Cepheids. Periods and light curves (Fig. 10.5) for the Cepheids provided material for identifying the types and using them for measuring the period-luminosity relation. A catalog of the variables, consisting of 82,000 individual measurements, was made available online. Mochejska et al. (2000) added a number of additional variable stars, for which magnitudes were determined using image subtraction, which was more efficient in detecting variables of small amplitude.

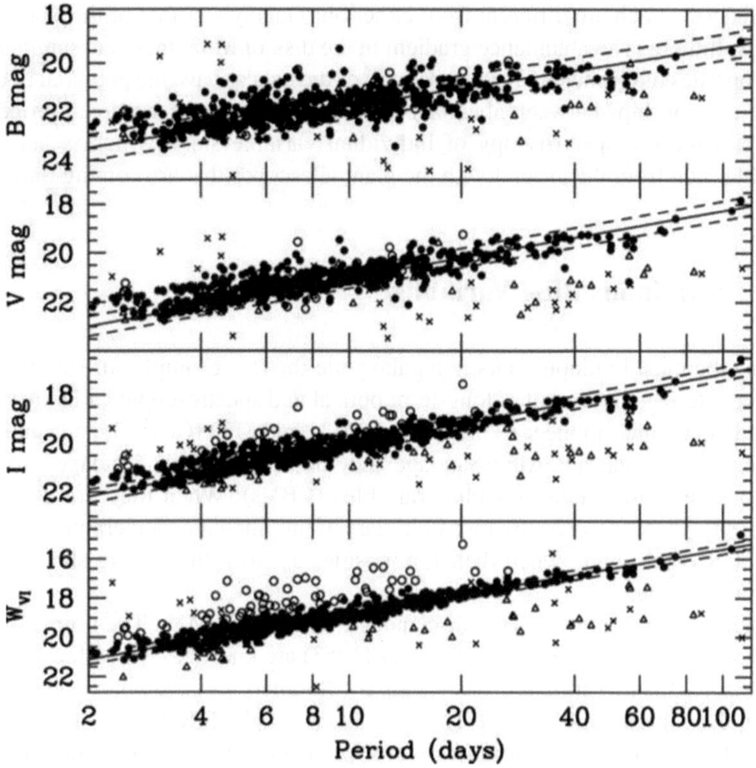

Fig. 10.6 Four versions of the period-luminosity diagram for M33. Shown from the *top down* are B, V, I and a reddening-free version. *Open circles* and *triangles*: possible blended images, crosses: poor light curves, solid line: best fit to the LMC P-L slope; dashed lines: 2 sigma deviation from fit (From Pellerin and Macri 2011). Used by permission, copyright AAS

More results based on the Macri et al. (2001b) survey of M33 were published by Pellerin and Macri (2011). The new results of the analysis of this large data set were the photometry of 564 Cepheid variables, as well as hundreds of other types of variable stars. The periods of the Cepheids ranged from 4 to 17 days, resulting in a well-filled P-L relationship (Fig. 10.6). Considerable attention was paid to photometric problems introduced by the dense background of the galaxy, which can lead to blending of images and overestimated magnitudes. This problem was attacked by examining Hubble Space Telescope images of available fields, which permitted de-blending in problem cases. This program is continuing and more results on the Cepheids and the other variables are expected in the next few years.

The future of the study of Cepheids in M33 will have to involve improvements in the photometry and in the extinction values for each Cepheid, which are presently uncertain at the 10% level. The usual goal, a precise P-L relationship, will be of value in determining both the distance to M33 and the effects of its heavy element abundances, which are different from either our Galaxy's or that of the Magellanic Clouds. If there is an abundance gradient in the disk of M33, then measurements of the properties of Cepheids in different parts of the galaxy have the potential of explicating relationships between abundances and measurable observational properties. This will require spectroscopy of individual variable stars, which is becoming possible even from the ground with the giant telescopes that are coming into use.

10.5 Luminous Blue Variables

Among the most luminous stars in a galaxy are the rare examples of variables that are characterized by irregular, long-term optical and spectroscopic variability. First noticed in Hubble's pioneer examination of M31 and M33 (Chap. 3) and studied by his then student assistant, Allan Sandage, they are called Hubble-Sandage Variables, or, more generally, luminous blue variables (LBVs). When they are at a high-luminosity phase their spectra resemble stars of intermediate temperature (late A or F spectral type), while when their luminosities are low, the spectra show Balmer lines in emission.

There are six confirmed Hubble-Sandage variables in M33. The four discussed in the original Hubble and Sandage paper (1953) are known as Variable 2, Variable A, Variable B, and Variable C (see Chap. 3). Another, Variable 83, was discovered by van den Bergh et al. (1975). The sixth LBV is called Romano's star (Romano 1978). Five additional candidates were found in 1996 by the examination of narrowband Hα images of M33 taken by Spiller with the Guillermo Haro Observatory in Mexico and examined by Corral (1996). These stars, which also appear in the catalog of Massey et al. (2007), need additional observations to be confirmed as members of the LBV class.

Hubble and Sandage (1953) were able to determine the characteristics of the light curves using a series of photographic plates taken at the Mt. Wilson Observatory. Nearly a dozen spectra of the variables were also available for examination at the time of their work and they were able to determine a relationship between luminosity and spectral properties. When the stars were seen near maximum luminosity hydrogen lines were seen in emission, some with P Cygni features. At the same time lines of ionized calcium were present. At minimum light the calcium lines were either not seen or faint and the Balmer hydrogen lines dominated the spectrum, together with helium lines in emission.

Our understanding of the LBV variables is not yet complete but recent research has produced some important information on the physical events leading to the remarkable changes detected in these objects. Already extremely luminous, the LBV variables sometimes increase in magnitude in an eruption of dust and gas and violent mass loss, resulting in a spectacular increase in luminosity. The Galactic object Eta Carina is an especially dramatic example of an LBV variable, probably similar to but more violent than the LBV variables in M33 (Humphreys et al. 2008).

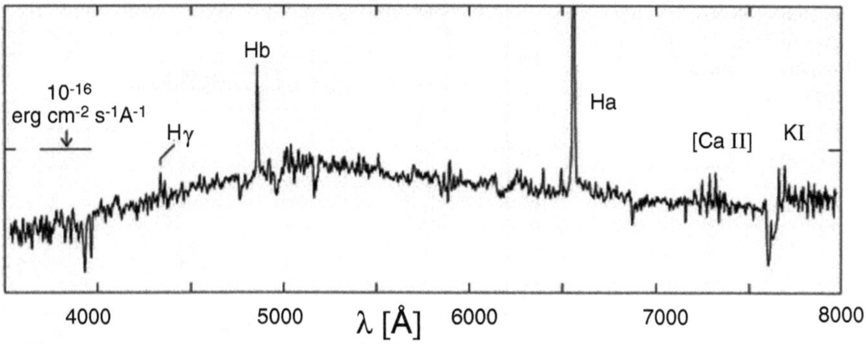

Fig. 10.7 The spectrum of Variable A, obtained in 2004 (From Humphreys et al. 2006). Used by permission, copyright AAS

Since the 1980s, considerable amounts of observational material have been collected and recently theoretical interpretation of the status of the stars has made progress. For example, Viotti et al. (2006) carried out a spectrophotometric study of four of the Hubble-Sandage variables in M33 and were able to derive explanations for the recent behavior of one of them. They found that Variable A was (in 2006) near minimum and its spectrum was solar-like but with strong hydrogen emission with the H-alpha line having an equivalent with of −310 Å. The bolometric luminosity was 6×10^6 times the solar luminosity. The star was interpreted as being an intermediate type supergiant surrounded by an expanding envelope of hydrogen and an obscuring dusty disk. Viotti et al. noted that at the time of their observations, Variable B was slowing changing color to the blue and Variable C was slowly fading and had a spectrum resembling that of Eta Carina.

Variable A was also the prime subject of a study by Humphreys and her colleagues (Humphreys et al. 2006), who had been observing its behavior for over three decades (Fig. 10.7). They attribute the changes in its spectrum to be caused by the ending of its eruptive phase, after almost 45 years, and a clearing of the cool, dusty envelope. A warm low density gas was visible in front of the star, curiously containing emission lines of [CaII] and KI (Fig. 10.8).

10.6 Long Period Variables

Long period variables (LPVs) of M33 became of interest in the 1980s, when studies of that type of star in the Magellanic Clouds demonstrated their potential both for understanding advanced stages of stellar evolution and for providing a distance indicator (Wood et al. 1983). The long period variables of the Magellanic Clouds showed a division into two types, highly luminous stars near the upper limit of the asymptotic giant branch and less luminous supergiants among the core-burning stars. The AGB stars showed larger amplitude in variation than the core-burning supergiants.

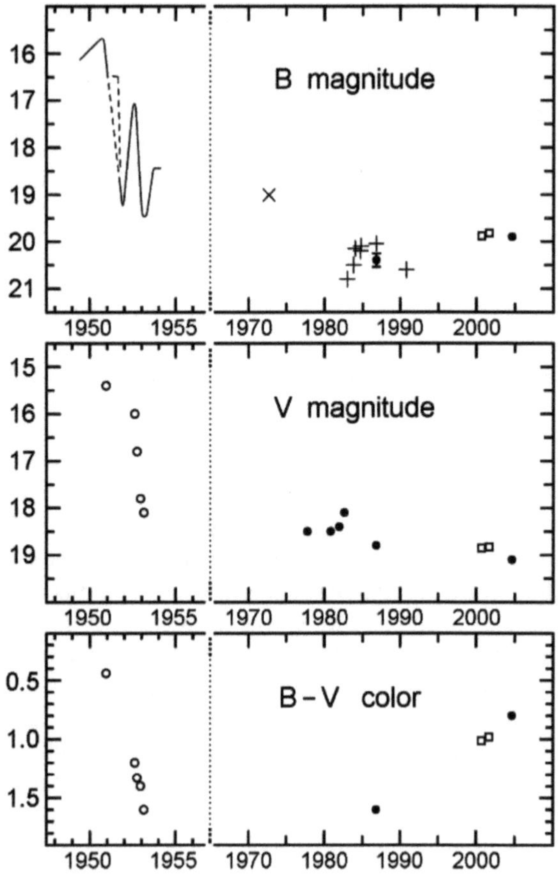

Fig. 10.8 The 50+ year light curves of Variable A (From Humphreys et al. 2006). Used by permission, copyright AAS

A search for LPVs in M33 was begun at the Kitt Peak Observatory in 1982 and carried out for 4 years (Kinman et al. 1987). An extensive series of photographic plates was taken with the 4-m telescope using red sensitive emulsions. Among the four different types of variable stars detected for which types and periods could be established were 54 LPV's.

Because the photographic data lacked calibration for faint stars in the red, infrared photometry of the LPVs in M33 was carried out with the 5-m telescope at the Palomar Observatory (Mould et al. 1990). The photometry showed that most of the variables were supergiants, only a minor number being asymptotic giant branch stars. Using H band photometry, which is relatively less subject to extinction corrections than visual photometry, permitted the construction of a period-luminosity diagram and a determination of a distance for M33. When the M33 data is compared to the PL relation for the Large Magellanic Cloud its distance modulus is found to be 24.64 ± 0.10, in reasonable agreement with the standard value (Chap. 1).

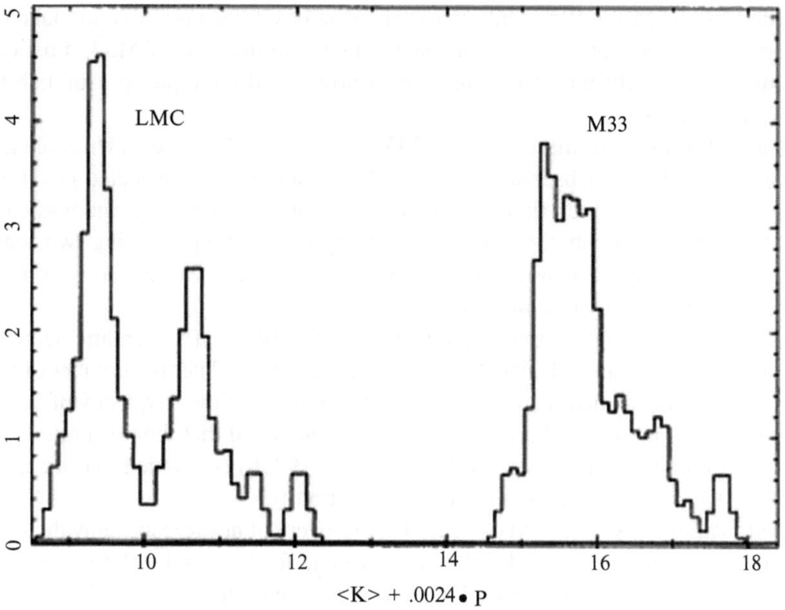

Fig. 10.9 The distribution of luminosities of long period variables in M33 and in the LMC (From Mould et al. 1990). The brightnesses are binned by $x = K + 0.0024P$, where K is the infrared K magnitude and the P is the Period. Used by permission, copyright AAS

Figure 10.9 shows a comparison of the luminosity distribution for the LPV's in the Large Magellanic Cloud and in M33. The ratio of the numbers of AGB variables, which occupy the fainter peak in the LMC distribution, to the supergiant type, which belong to the higher peak at brighter magnitudes, is very different for the two galaxies. However, it is not clear whether this difference reflects the true ratio of types or is the result of incomplete detection of the fainter AGB stars.

A new search for long period variables in M33 was carried out from 1996 to 2003, using data obtained with the Whipple Observatory 1.2 m telescope and the WIYN telescope on Kitt Peak (Pellerin and Macri 2011). The former observations were part of the DIRECT project, which was designed to measure accurate distances to nearby galaxies and the latter observations were obtained to provide time coverage for more accurately determining periods of some of the longer period variables. These data provided information on more than 400 LPVs.

10.7 Novae

Several hundred novae have been observed in our Galaxy and it is estimated that at least 30 nova outbursts occur annually. The study of novae in other galaxies has had as its primary goal a determination of the global rate of occurrence, so as to compare it to that for the Milky Way. Only in recent years has it been possible to make more astrophysical observations of extragalactic novae. In the case of M33, most of the literature deals only with the detection of novae and attempts to establish their rate of recurrence.

The first novae to be discovered in M33 were noted by Hubble and his colleagues at the Mount Wilson Observatory (Chap. 3). A total of five suspected novae were known by 1956, at which time a systematic attempt to discover novae was carried out at Mt. Wilson by photographing M33 every possible night during two seasons (Arp 1956). Arp found no novae and concluded that novae are rare in M33 compared to the rate observed by him in M 31.

In 1963 the Asiago Observatory near Bologna began a systematic search for novae in M33 (Rosino and Bianchini, 1973). Twenty years later the number of novae and candidate novae had increased to 14. A calculation of the frequency of novae in the galaxy was made by Sharov (1983a), who concluded that the nova rate in M33 is approximately one per year. He later (Sharov 1983b) concluded that the correct figure is 0.45 novae per year, with a large uncertainty.

Della Valle (1988) analyzed the record of discovery of novae, especially the Asiago Observatory data, and concluded that the nova rate for M33 is 4 ± 2 novae per year. He later (Della Valle 1992) revised the figure, having re-analyzed the time coverage and its incompleteness, together with other systematics in the data, and concluded that the rate is 4.7 ± 1.5 novae per year.

A much more complete survey, using Halpha images, provided better statistics on the annual nova rate (Williams and Shafter 2004). The 8 year campaign used the Mount Laguna Observatory 1 m telescope and consisted of five different fields, 14' × 14'in size, which covered approximately 60% of the disk of M33. During that period six novae were detected, which implies a nova rate of 2.5 (±0.3) per year. This value is roughly halfway between the extremes of 4.7 (Della Valle 1992) and 0.45 (Sharov 1993). When the number is adjusted to give the number per unit mass of the host galaxy, the result is a luminosity-specific rate of 2.34×10^{-10} solar luminosities per year. This value, within the uncertainties, agrees with luminosity-specific rates for other galaxies, possibly with the exception of the Magellanic Clouds, which appear to have unusually high values.

By the beginning of the twenty-first century the field of extragalactic nova astronomy had expanded to include spectroscopic observations and information about the physical properties of the novae. An example is a photometric and spectroscopic study of the nova M33 2007 No. 1 (Shafter et al. 2007). Using the Hobby-Eberly Telescope in Texas, the astronomers concluded that the nova belongs to a special class called He/N novae (Fig. 10.10). Broad lines of the hydrogen Balmer series, indicating velocities of approximately 5,500 km/s were detected. Strong emission lines of NII, NIII, HeI and HeII were also evident. This particular nova faded rapidly, as is common for Galactic novae of this type. Photometry of the nova suggested that its absolute magnitude was at least as bright as −8.2.

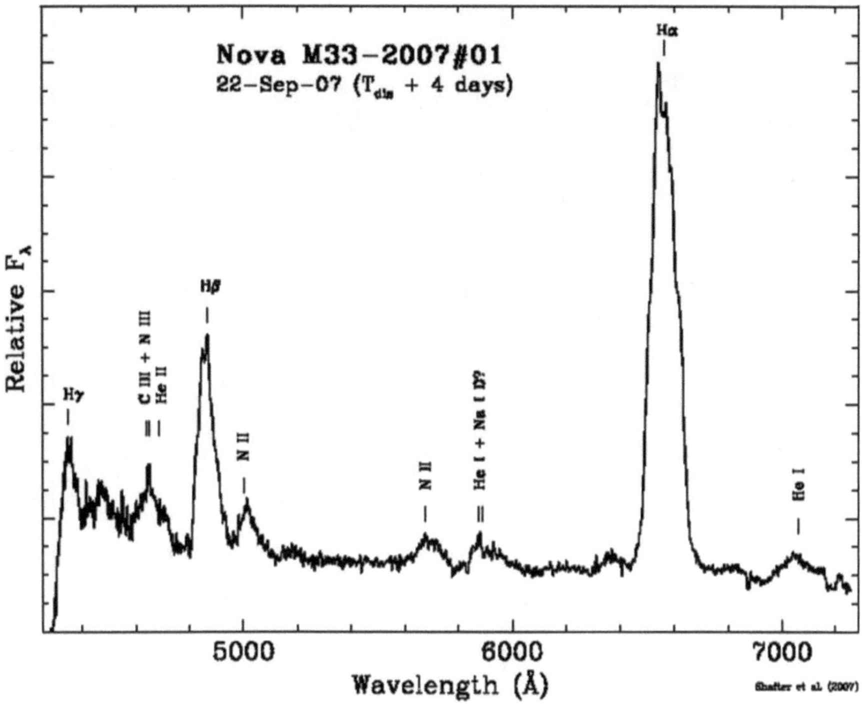

Fig. 10.10 A spectrum of a nova in M33 observed with the Hobby-Eberle Telescope. Prominent emission lines are identified (From Shafter et al. 2007). Used by permission

10.8 RR Lyrae Variables

The first report of the presence of RR Lyrae variables in M33 was published by Pritchet (1988), but details have not been published. Almost 20 years later Sarajedini et al. (2006) published a detailed description of an HST survey of two fields in M33, both near clusters that had been described in other papers (see Chap. 6). Seventy-two RR Lyrae variables were identified, of which 64 were type ab. Periods ranged from ~0.8 to ~0.3 days (Fig. 10.11). The shortest period variables make up a population that is not found in Galactic globular clusters; they are probably old but metal-rich, a combination that is not found in our globular clusters. That explanation is supported by the finding of an old, metal-rich component among the M33 halo stars (Chaps. 9 and 10).

Using the relationship between metallicity and period of Galactic RR Lyraes, the authors determined the metallicities of the M33 sample (Fig. 10.12). There is a wide range, from [Fe/H] of 0.5 to –2.0, with a mean of approximately –1. The data suggest a possible bimodal distribution, with one peak at –0.7 and another at –1.3. The former agrees well with the metallicities of disk stars (Chap. 8) and the latter with the abundances of the old halo field stars and globular clusters (Chap. 6).

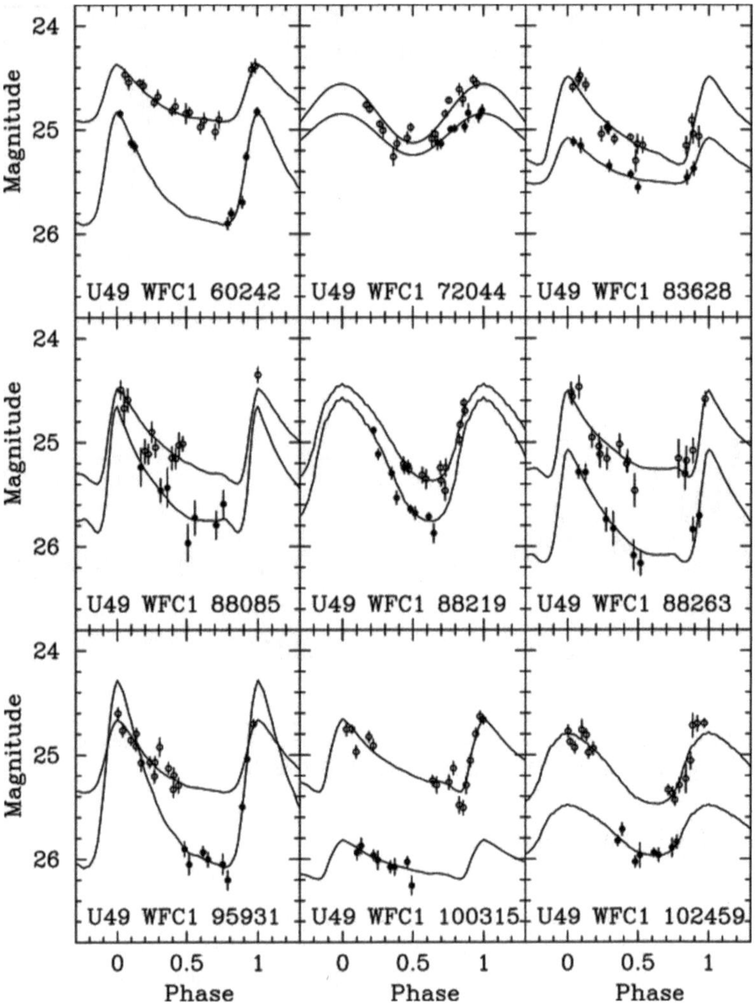

Fig. 10.11 Light curves for RR Lyrae variables in M33. *Open circles* are I data and *filled circles* are V data (From Sarajedini et al. 2006). Used by permission, copyright AAS

Traditionally the brightnesses of RR Lyrae variables have been one of the most reliable methods for determining distances to old populations in galaxies. Sarajedini et al. (2006) determined the distance to M33 in this way. They used the following relation between metallicity and absolute magnitude

$$M_V = 0.23[Fe/H] + 0.93$$

which leads to a determination of the distance modulus of 24.67 ± 0.08 magnitudes, corrected for reddening. This value happens to be exactly the value adopted for this book (Chap. 1), which was a mean of a large number of recent determinations.

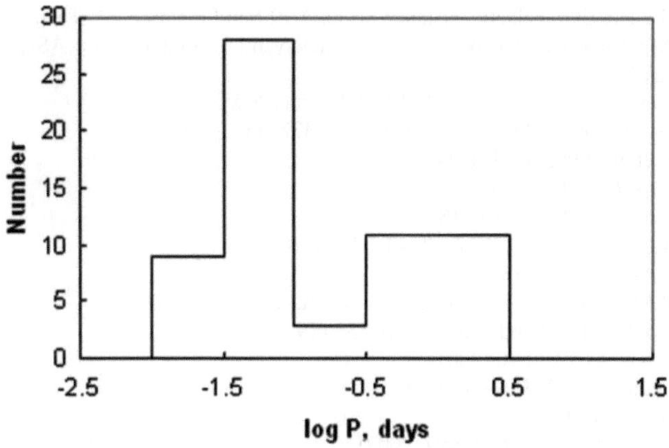

Fig. 10.12 The distribution of derived metallicities of the ab type RR variables in M33 (Based on data from Sarajedini et al. 2006). Used by permission, copyright AAS

10.9 Summing Up

After being neglected for decades following Hubble's work, the variable stars of M33 recently have drawn significant attention, especially in the last 20 or so years. Even eclipsing variables have become important, especially for establishing a geometrical distance to the galaxy. The first such attempt has now been made, with a result that differs by 13% from the photometric distance, a disagreement that is worrying, but that needs more examples before a concern is raised. The most important variables are the Cepheids, which not only provide a distance but also allow an examination of the differences that are caused by the chemical abundances in them, which are intermediate between that of Cepheids in the Galaxy and those in the Magellanic Clouds. Long period variables are found to be very abundant in M33 and a beginning of the study of this rich resource has been made. The Hubble-Sandage variables, or LBVs, in M33 are being followed carefully, as they are a significant fraction of this kind of explosive star known in the Local Group. Lastly, the RR Lyrae variables are finally being studied, now that the HST has made it possible. They show some interesting differences from those in the Galaxy, especially in our old globular clusters. Tentatively, the differences may be due to the presence of a population of intermediate metallicity RR Lyraes, rare in other galactic environments so far studied.

References

Arp, H.C.: AJ **61**, 15 (1956)
Cioni, M.-R.L., et al.: A&A **131**, 146 (2008)
Corral, L.J.: AJ **112**, 1450 (1996)

Della Valle, M.: The Extragalactic Distance Scale. ASP, San Francisco (1988). p. 73
Della Valle, M.: Vina del Mar Workshop on Cataclysmic Variable Stars. ASP, San Francisco (1992). p. 292
Dubus, G., Long, K.S., Charles, P.A.: ApJ **519**, L135 (1999)
Freedman, W.L., Wilson, C.D., Madore, B.F.: ApJ **372**, 455 (1991)
Hartman, J.D., et al.: MNRAS **371**, 1405 (2006)
Hubble, E.P.: ApJ **69**, 103 (1929)
Hubble, E.P., Sandage, A.R.: ApJ **118**, 353 (1953)
Humphreys, R.M., Sandage, A.R.: ApJS **44**, 319 (1980)
Humphreys, R.M., et al.: AJ **131**, 2105 (2006)
Humphreys, R.M., Davidson, K., Koppelman, M.: AJ **135**, 1249 (2008)
Kinman, T., Mould, J.R., Wood, P.R.: AJ **93**, 833 (1987)
Macri, L.M., et al.: AJ **121**, 870 (2001)
Madore, B.F., et al.: ApJ **294**, 560 (1985)
Massey, P. et al.: AJ **134**, 247 (2007)
McQuinn, K.B.W., et al.: ApJ **664**, 850 (2007)
Mochejska, B., et al.: BAAS **32**, 1576 (2000)
Mould, J., et al.: ApJ **349**, 503 (1990)
Pellerin, A., Macri, L.M.: ApJS **193**, 26 (2011)
Pritchet, C. R.: in "The Extragalactic Distance Scale", ASP. **59** (1988)
Romano, G.: A&A **67**, 291 (1978)
Rosino, L., Bianchini, L.: A&A **22**, 461 (1973)
Sandage, A.R.: AJ **88**, 1108 (1983)
Sandage, A.R., Carlson, G.: ApJ **267**, L25 (1983)
Sarajedini, A., et al.: AJ **132**, 1361 (2006)
Shafter, A. W., et al.: Astron Telegrams No. 1225 (2007)
Sharov, A.S.: Sov. Ast. Letters **9**, 253 (1983a)
Sharov, A.S.: Sov. Ast. Letters **9**, 340 (1983b)
Sharov, A. S.: Sov. Ast. Letters, **19**, 147 (1993)
van den Bergh, S., Herbst, E., Kowal, C.T.: ApJS **29**, 303 (1975)
Viotti, R.F., et al.: A&A **458**, 225 (2006)
Williams, S.J., Shafter, A.W.: ApJ **612**, 867 (2004)
Wood, P.R., Bessell, M.S., Fox, M.W.: ApJ **272**, 99 (1983)

Chapter 11
X-Ray Sources

11.1 Early Surveys

X-rays from M33 were first detected in 1981, when Long et al. (1981) obtained observations with the Imaging Proportional Counter on the Einstein Observatory.

They identified about a dozen point sources, which had X-ray luminosities ranging from 10^{37} ergs/s to 10^{39} ergs/s, if at the distance of M33. One source stood out, making up about 3/4ths the total detected X-ray flux. It appeared to be located at the nucleus of M33 and was the brightest steady X-ray source in the Local Group. During the Einstein decade a total of 17 sources was identified, including one background and one foreground source. Unresolved emission was also detected from the M33 disk.

ROSAT observations in the 1990s (Schulman and Bregma 1995; Long et al. 1996) increased the number of known X-ray sources to 57 and revealed that the diffuse emission appeared to trace the optical structure, including the more prominent arms near the nucleus (Fig. 11.1). Eventually the ROSAT observations detected 184 sources within 12 kpc of the nucleus (Haberl and Pietsch 2001). Some of these, of course, were background and foreground sources.

The next stage in the identification of M33 candidate X-ray sources was introduced by the XMM-Newton telescope, which obtained a deep survey of the main body of M33 (Pietsch et al. 2004; Misanovic et al. 2006). A total of 447 sources was identified within the D_{25} optical isophote and a partial separation of M33 sources from foreground and background sources was made. The nature of some of the M33 sources could be learned by comparing the X-ray characteristics, particularly X-ray hardness ratios and positions, with catalogs of objects based on surveys at other wavelengths. In this way several candidate SN remnants and X-ray binaries were identified. As well, there were several supersoft sources. Among the several transient sources were two optical novae. The high density of sources and the bright X-ray diffuse emission in the central regions made it difficult to disentangle and characterize all of the sources at the resolution of the surveys made prior to the launch of the Chandra Observatory.

P. Hodge, *The Spiral Galaxy M33*, Astrophysics and Space Science Library 379,
DOI 10.1007/978-94-007-2025-1_11,

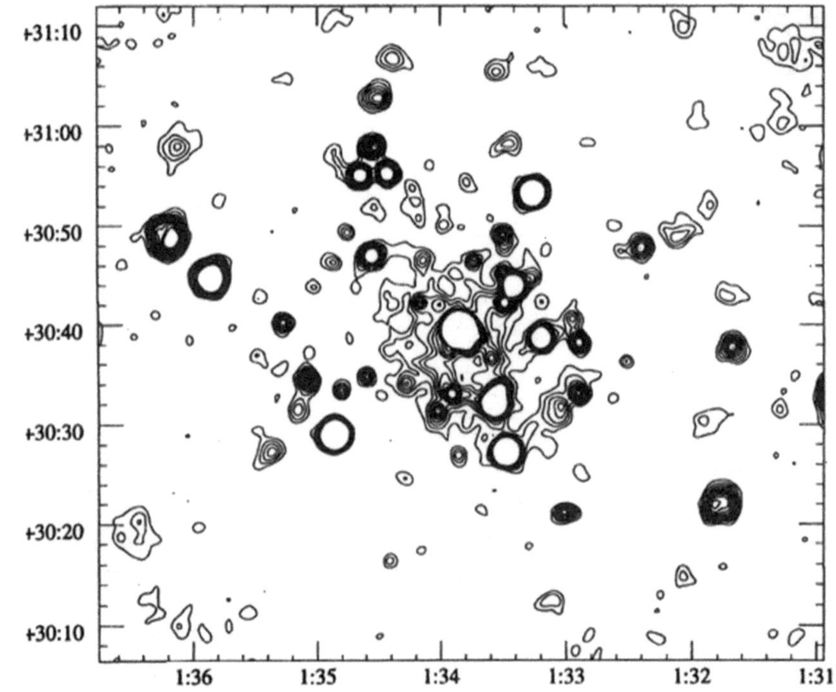

Fig. 11.1 A map of X-ray sources in M33 based on ROSAT observations. The area is approximately 1° square (From Long et al. 1996). Used by permission, copyright AAS

11.2 The Chandra Survey

The Chandra Observatory was launched on July 23, 1999 by the Space Shuttle Columbia. One of NASA's Great Observatories, it was the most powerful X-ray telescope so far placed in orbit. One of its prime targets was the central area of M33, for which previous telescopes were unable to resolve the complicated X-ray emission. A comprehensive program for this purpose, called ChASeM33 (for Chandra ACIS Survey of M33, where ACIS stands for the Advanced CCD Imaging Spectrometer) was carried out by an international team of 12 astronomers (Plucinsky et al. 2008). The observations were begun in September, 2005 and they covered the inner regions of M33 out to a distance from the nucleus of 4 kpc.

Figure 11.2 shows a mosaic image of M33 constructed from the ChSSeM33 observations. The dominant feature is the nucleus, which is surrounded by an area of diffuse emission, which extends out to approximately 10 arc minutes from the center. Discrete sources throughout the region were detected and classified according to their hardness (ratio of fluxes at different wavelengths). A total of 394 sources was identified in the "first-look" survey. The brightest discrete objects included well-known features, such as the giant HII region NGC 604, and bright X-ray

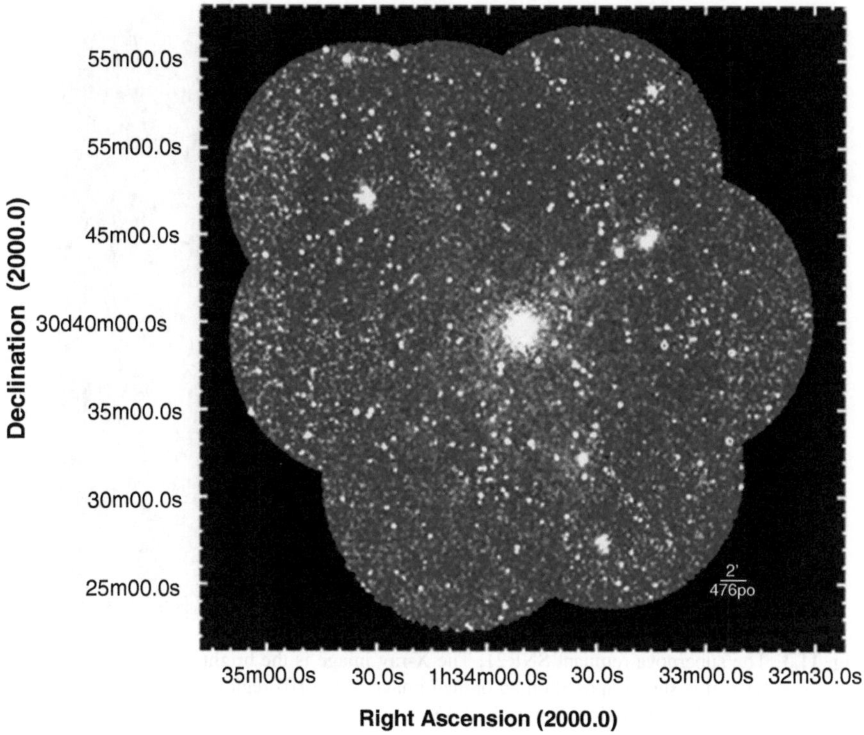

Fig. 11.2 A mosaic X-ray image of M33 based on the Chandra first-look survey of the galaxy's central area (From Plucinsky et al. 2008). Used by permission, copyright AAS

binaries, such as M33 X-7. Also bright were Galactic foreground stars and some background AGN, types that were characterized by large hardness ratios. The foreground and background sources were thought to make up at least half of the catalog and definite identification of them will require more study, including detailed optical observations.

11.3 Supernova Remnants

It was expected that many of the soft sources (with low hardness ratios) would be supernova remnants. Of the 100 optical SNRs cataloged in M33 (Gordon et al. 1998; Ghavamian et al. 2005), the Chandra survey found coincidences in position with 31. A particularly interesting case is SNR21, which was detected by Chandra as an elliptically-shaped shell imbedded in the giant HII region NGC 592 (Gaetz et al. 2007, Fig. 11.3).

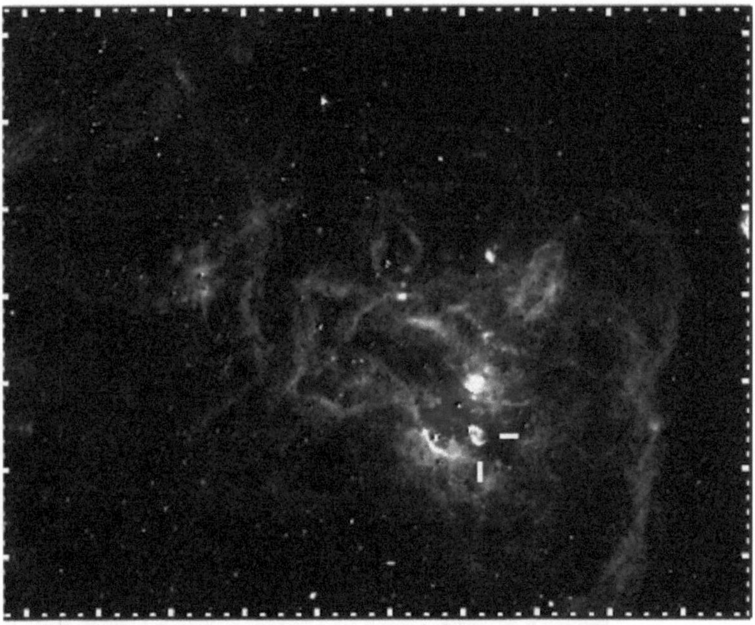

Fig. 11.3 The supernova remnant SNR21. The X-ray image is the bright structure indicated by two *white bars*. It is superimposed on an optical image of the HII region NGC 592 (From Gaetz et al. 2007). Used by permission, copyright AAS

11.4 Foreground Sources

There are large numbers of Galactic stars superimposed upon the face of M33; the US Naval Observatory star catalog includes nearly 9,000 stars within the ChASeM33 coverage. Identifying these foreground stars among the sources in the Chandra catalog has required considerable effort, which has been aided by the surveys of M33 stars at optical and uv wavelengths published by Massey et al. (1996, 2006). The uv-bright stars include O and B supergiants and WR stars and they lie preferentially along spiral arm segments. A total of 10 of the Chandra sources lie close (within 5 arcsec) to Massey stars, while only one would be expected by chance. Thus at least a few Chandra sources are known high-temperature Galactic foreground stars. Additionally, there is an M-type foreground source (ChASeM33 #175) that was identified by the presence of X-ray flares.

11.5 NGC 604 as an X-Ray Source

The giant HII region NGC 604 is one of the most luminous and massive HII regions in the Local Group (Skelton 1999). The Chandra X-ray image showed a bright diffuse source and a point-like source (Fig. 11.4). The spectrum of the diffuse emission was

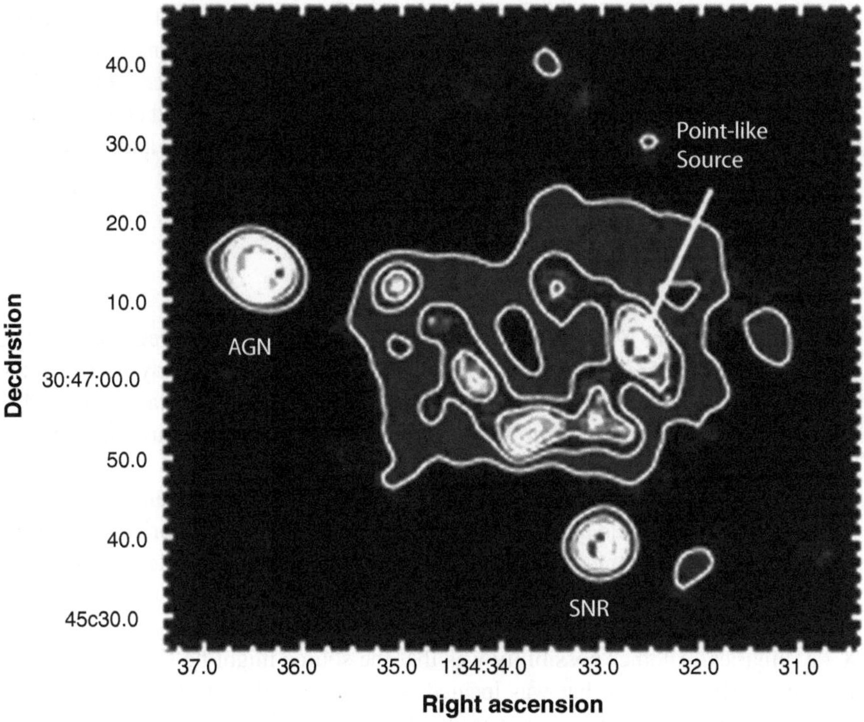

Fig. 11.4 A Chandra X-ray image of the HII region NGC 604. The point-like source in the nebula is indicated. The two objects outside the HII region are unrelated sources (From the "first look" paper, Plucinsky et al. 2008). Used by permission, copyright AAS

well-fit to a thermal emission model with a temperature of 0.2 keV and a flux of 5×10^{-14} ergs cm^{-2} s^{-1}. The point source was too weak to determine its nature with certainty.

Two other bright sources are shown in Fig. 11.4, both apparently unrelated to NGC 604. Below the HII region there is a cataloged SNR, GKL98-94. Its X-ray spectrum was fit with a thermal model with $N_H = 0.56\times10^{22}$ cm^{-2}, kT = 0.25 keV and F_X (0.5–2 keV) = 2.5×10^{-15} ergs cm^{-2} s^{-1}. The point source to the east (left) of NGC 604 has the characteristics of a background AGN. Its spectrum is harder than that of the other sources in this region.

11.6 The Southern Arm

One of the best-defined and richest sections of M33's spiral arm structure is located about 6 arc minutes south of the nucleus. Here is the maximum in the star formation rate in the galaxy and, according to the optical papers cited above, it also is the location of an excess of SNRs. There are 26 optically-identified SNRs located within the ChASeM33 area and 10 have clearly detected X-ray counterparts. Many of the others have marginal detections and may also be X-ray sources. Figure 11.5 demonstrates the fact that the arm is rich in both X-ray sources and star formation activity and shows some clear coincidences between X-ray sources and SNRs.

11.7 X-Ray Binaries

There are two bright X-ray binaries in M33 and both of them are remarkable. The most luminous is M33 X-8 which is the most luminous steady X-ray source in the Local Group of Galaxies, outshining all others, even those in the giant spirals M31 and the MWG. The other bright X-ray binary, M33 X-7, is also unusual, as it is an eclipsing X-ray pulsar.

M33 X-8 was first detected by Long et al. (1981) using the Einstein Observatory. It was seen to be coincident with the nucleus of M33 and to comprise 70% of the entire galaxy's X-ray emission. Its luminosity in the energy range 0.15–4.5 keV is a whopping 10^{39} ergs sec^{-1}. For many years the nature of X-8 was unknown. Possibilities considered were that it might be small AGN or a collection of normal X-ray binaries. Another possibility was that the source might be the hot giant star with an O9III spectrum that was located one arc second away from the nucleus. Other pre-Chandra observations were made by ROSAT, covering a 6-year period (Dubus et al. 1997), which showed a variation in brightness with a 106 day cycle. The authors argued that the periodicity and amplitude implied that the source is a single physical object and they put forward the idea that it might be a binary consisting of a black hole and a ten solar mass giant star. Chandra observations showed that the X-ray source is positionally coincident with the M33 nucleus and eliminated the possibility that the nearby bright blue star was the source (Dubus and Rutledge 2002). The details of the X-ray spectrum and the characteristics of its variability were found to resemble GR 1915 + 105, a Milky Way micro-quasar, which has a 14 solar mass black hole that is accreting material from a K-type giant companion. The main difference shown by M33 X-8 is its high X-ray luminosity, which may be the result of a particularly favorable viewing angle.

M33 X-7 is also a remarkable object. Observations with the Einstein Observatory (Long et al. 1981) were the first to detect it and subsequent observations showed it to be variable in luminosity. Peres et al. (1989) proposed that it varies with a short period, which was later shown to be 3.45 days (Larson and Schulman 1997; Dubus et al. 1999).

The characteristics of M33 X-7 are best understood if it is a high-mass eclipsing binary, consisting of a neutron star and a supergiant star with a mass of 15–49 solar masses. The separation of the two objects is calculated to be 25–33 solar radii. Furthermore, ROSAT observations detected a 0.31s pulse period, indicating that the neutron star is a pulsar. Thus X-7 is an interesting case of an X-ray neutron star high mass binary that is also a pulsar.

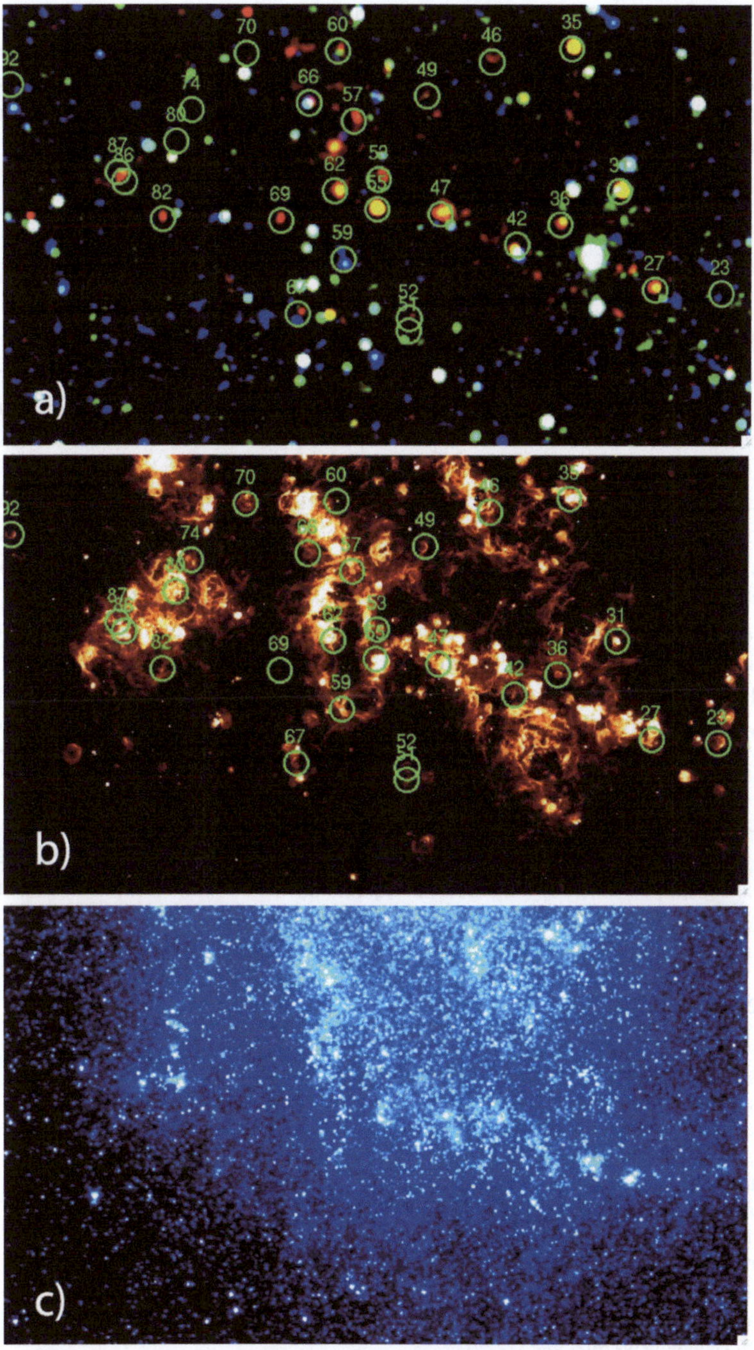

Fig. 11.5 The Southern Arm of M33. Panel (**a**) is the Chandra map showing the X-ray sources and panel (**b**) is an Hα image from the LGGS (Massey et al. 2006). The positions of SNRs are indicated with their GKL98 catalog numbers. (c) is an optical image (From Plucinsky et al. (2008), where the published version is color-coded). Used by permission, copyright AAS

In 2009 another X-ray binary, M33 X47, was discovered by the CHANDRA ACIS survey (Pietsch et al. 2009). It has a period of 1.732479 days. It had a hard X-ray spectrum and a luminosity of 2×10^{37} ergs/s during its high state. The data fit best for a binary consisting of a neutron star and an 11 solar mass hot supergiant companion.

11.8 Summing Up

Starting with the Einstein maps of the 1980s, the X-ray view of M33 had three important components: (1) a very luminous source that was apparently at the center of M33, (2) several fainter sources scattered over the disk, and (3) a faint diffuse component. As the observations improved, first with ROSAT and then with XXM-Newton, the number of discrete sources increased to several hundred. With the launching of Chandra it became possible more clearly to separate sources into their different natures: some were X-ray binaries, some were SNRs and some were super-soft sources (and some, of course, were foreground and background objects). The supernova remnants included SNR21, a shell located in the giant HII region NGC 592. Another giant HII region, NGC 604, was found to have a bright point source near its center surrounded by an irregular cloud of unresolved X-rays. The galaxy's spiral arms were seen to be sprinkled with fainter X-ray sources, probably mostly X-ray binaries and SNRs. This description so far could be that of any small spiral galaxy. Where M33 stands out is in its two brightest point sources, M33 X-7 and M33 X-8. The former turns out to be an eclipsing binary, with a period of 3.45 days. The components are a pulsar and a massive supergiant. M33 X8 was found to be even more remarkable: it is probably a micro-quasar at the center of the galaxy, consisting of a 14 solar mass black hole, possibly a member of a massive binary.

References

Dubus, G., Rutledge, R.E.: MNRAS **336**, 901 (2002)
Dubus, G., et al.: ApJ **490**, L47 (1997)
Dubus, G., et al.: MNRAS **302**, 731 (1999)
Gaetz, T.J., et al.: ApJ **663**, 234 (2007)
Ghavamian, P., et al.: AJ **130**, 539 (2005)
Gordon, S.M., et al.: ApJS **117**, 89 (1998)
Haberl, F., Pietsch, W.: A&A **373**, 438 (2001)
Larson, D.T., Schulman, E.: AJ **113**, 618 (1997)
Long, K.S., et al.: ApJ **246**, L61 (1981)
Long, K.S., et al.: ApJ **466**, L750 (1996)
Massey, P., et al.: ApJ **469**, 629 (1996)
Massey, P., et al.: AJ **131**, 2478 (2006)
Misanovic, Z., et al.: A&A **448**, 1247 (2006)
Peres, G., et al.: ApJ **336**, 140 (1989)

Pietsch, W., et al.: A&A **426**, 11 (2004)
Pietsch, W., et al.: ApJ **694**, 449 (2009)
Plucinsky, P.P., et al.: ApJS **174**, 366 (2008)
Schulman, E., Bregma, J.N.: ApJ **441**, 568 (1995)
Skelton, B.: BAAS **31**, 1509 (1999)

Chapter 12
Dynamics and Mass

12.1 The HII Region Velocity Curve

The first relatively complete study of the dynamics of M33 was carried out by Mayall and Aller (1942). Using spectra taken at the Lick Observatory, they obtained velocities of 25 emission nebulae in M33, selected from objects ranging in position from near the center to the outer parts of the visible disk. Using the mean velocity of a set of the objects close to the center they determined that the systemic velocity of M33 is -167 ± 5 km/s. Using the inclination to the line of sight value obtained from the ellipticity of the image, they were able to produce a diagram showing the relationship between distance from the center of the galaxy and the radial velocity of the source (Fig. 12.1). The velocities had relatively large uncertainties because of the low dispersion of the spectra, which was approximately 300 Å/mm in the blue. Nevertheless, the results illustrated the fact that the galaxy rotates with velocity increasing with distance from the center out to distances of about 30 arcmin.

A mean rotation curve for M33 was derived by folding the two halves of the major axis together to provide a curve of the absolute value of the velocity versus distance, with the different measurements weighted according to their precision. Figure 12.2 shows the result, in which the observations have been combined into six bins and fit by eye with a smooth curve. The interpretation of this rotation curve was that M33 consisted of two components: a central body that rotates like a solid body and an outer disk that shows Keplerian rotation. This interpretation was adopted by most succeeding studies of the velocity curve for M33 (and other spiral galaxies) until the 1960s, when it was found to be wrong, for surprising reasons.

In order to determine the distribution of mass and the total mass of M33, Wyse and Mayall (1942) derived a simple model of a thin-disk galaxy that had a rotation curve similar to that derived by Mayall and Aller (1942). Models of circularly symmetric thin disks with density distributions showing various relationships with distance from the center were constructed to determine the force curve in the plane. Assuming simple circular motion, it was possible to convert the force curve into a predicted rotation curve. Using the observations, Wyse and Mayall were able to find

P. Hodge, *The Spiral Galaxy M33*, Astrophysics and Space Science Library 379,
DOI 10.1007/978-94-007-2025-1_12, © Springer Science+Business Media B.V. 2012

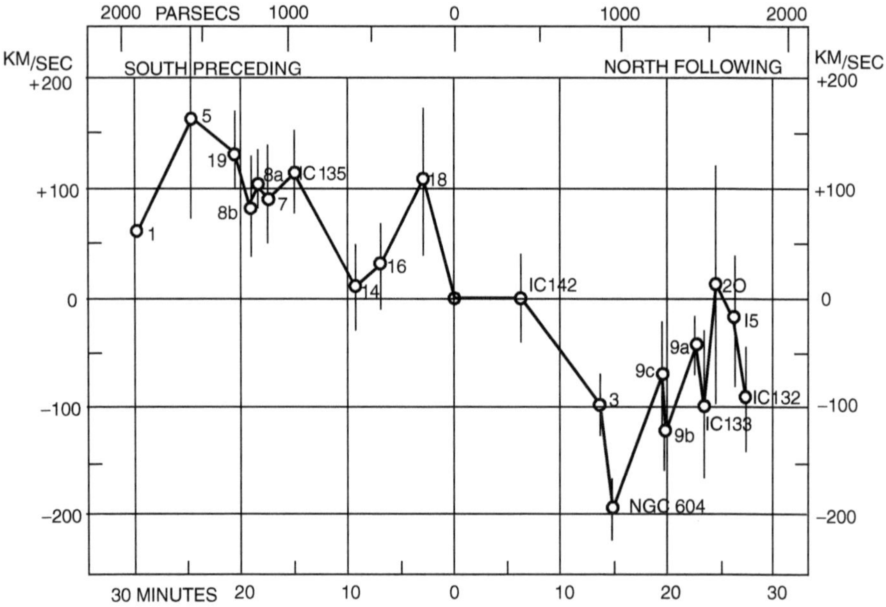

Fig. 12.1 The radial velocities of HII regions in M33 determined by Mayall and Aller (1942). Used by permission, copyright AAS

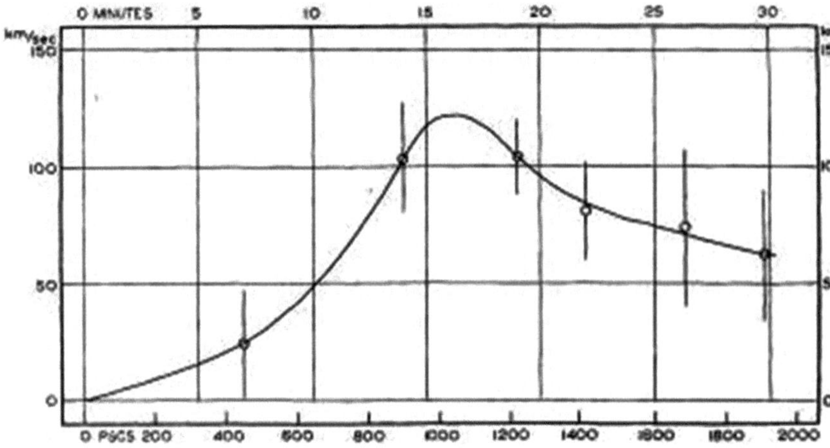

Fig. 12.2 The relationship between the rotational velocities and distance from the center of M33 (Mayall and Aller 1942). Used by permission, copyright AAS

a projected density distribution that fit the observed velocity curve (Fig. 12.3). The average projected density in the disk of M33 was found to be 500 solar masses per square parsec and the total mass of M33 was calculated to be 1.7 billion solar masses (note that Wyse and Mayall used Hubble's distance to M33, namely 220 kpc, which is a factor of 3.86 smaller than the currently accepted value).

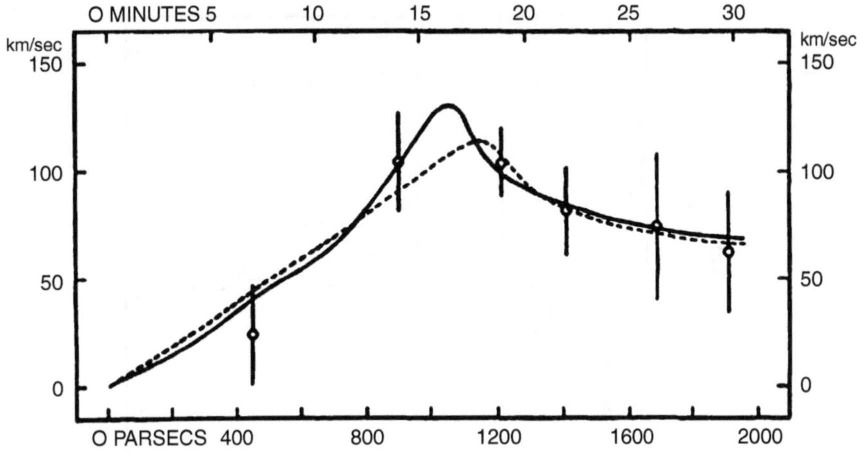

Fig. 12.3 Wyse and Mayall's (1942) fits to the rotation curve of M33. Used by permission, copyright AAS

During the late 1950s and 1960s considerable progress was made in determining the masses of galaxies using long slit spectra to record HII region velocities across galaxies' disks (Burbidge and Burbidge 1960, and many following papers). Dozens of spiral galaxies were observed and their HII regions' velocities reached out to near the visible edge of the disks. The mass distribution models that were used were sophisticated and provided good fits to the data. However, the data at the largest distances from the centers of the galaxies were few and sparse, so that the Keplerian portion of the rotation curve was not well-represented.

It was 20 years before M33's rotation curve was re-examined, when Brandt (1965) constructed a more sophisticated model based on new observational data, including new optical spectra taken with the Mount Wilson 100-inch telescope and radio observations of the H I velocities, which extended the rotation curve out to a degree from the center. The total mass, calculated from fitting realistic three-dimensional models of the disk, was found to be 20 billion solar masses.

When reviewing the values of the mass of M33 mentioned above one should remember that the distance to the galaxy was not well established. Beginning with Hubble's determination, the distance was underestimated and thus the masses were also underestimated. Furthermore, as will be discussed below, the rotation curves were unreliable, especially in the outer areas, where the radial velocities were also underestimated.

A different kind of survey was the extensive H-alpha interferometric study by Carranza et al. (1973). With a total of 1,048 velocities across the disk it was possible to investigate the rotation curve in detail and to detect the effects of the mass concentrations in the spiral arms. The rotation curve in M33 was seen to agree well with Brandt's (1965) models. However, it did not show a downturn, even at a distance from the center of 6 kpc. There was no evidence of Keplerian motion in the outer regions.

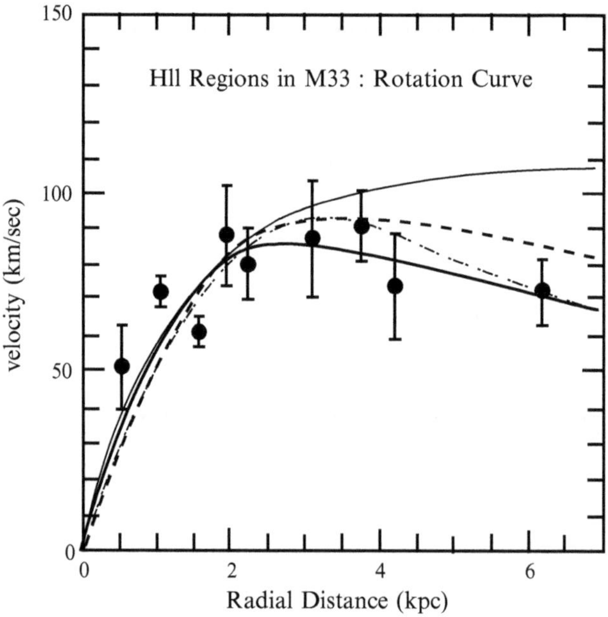

Fig. 12.4 The rotation curve obtained by Zaritsky et al. (1989). The *dots* are the measurements with their errors, the upper, *dotted curve* is the radio HI curve (Rogstad et al. 1976) and the *other curves* are for various assumed mass distributions. Used by permission, copyright AAS

Nevertheless, the question remained open: was the rotation curve being correctly made and interpreted? Using spectra of HII regions taken with the Steward Observatory 90-inch telescope, Zaritsky et al. (1989) measured the radial velocities of 55 HII regions across the disk of M33. Their rotation curve (Fig. 12.4) was similar to previous results and showed a downturn after about 4 kpc, with the outermost value about 10 km/s lower than the peak value at 3 kpc. Three different models for the mass distribution were derived and the preferred one had a mass-to-light ratio (in solar units) that is approximately constant with a value of 2.1.

12.2 The HI Velocity Curve

Neutral hydrogen velocities in M33 provided much more complete data for the construction of rotation curves than the optical data. Volders (1959), Dieter (1962) and others in the 1960s and early 1970s mapped the HI 21-cm emission from M33, but these observations had low resolution because of the relatively small size of the radio telescopes used. Better resolution was achieved by the use of the Cambridge Half-Mile Telescope (Wright et al. 1972; Warner et al. 1973), which had a spatial resolution of 1.5 by 3 arcmin and a velocity resolution of 39 km/s. Their rotation

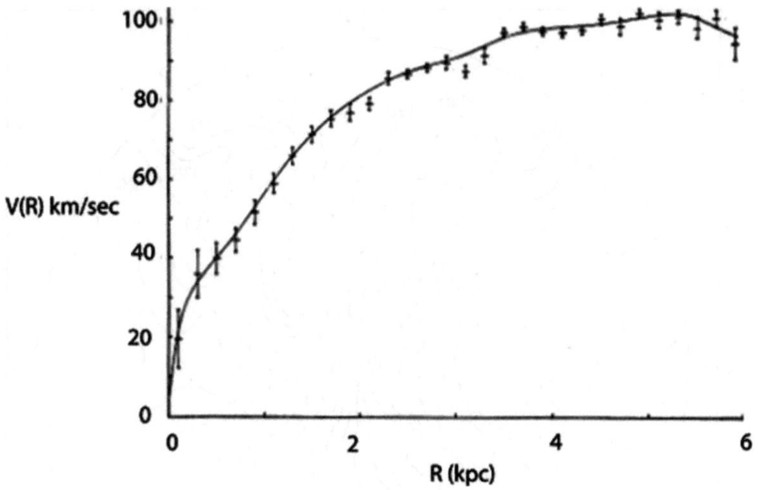

Fig. 12.5 The HI rotation curve for M33 (Warner et al. 1973). Copyright RAS

curve showed an increase with radius out to about 5 kpc (using our adopted distance of 850 kpc), at which position the velocity was about 100 km/s. Beyond 5 kpc the curve was nearly level until at 7 kpc it turned down to a velocity of 70 km/s (Fig. 12.5). The inferred total mass out to 7 kpc was found to be 9.2 billion solar masses.

A slightly different fit to the HI data was published by Lohmann (1974), who showed a curve with a definite downturn starting at 3 kpc (by our distance). The outer parts of his curve seemed to show the expected Keplerian shape, though it did not agree very well with the data.

Aperture synthesis observations of HI in M33 were carried out a few years later by Rogstad et al. (1976) at the Owens Valley Radio Observatory. The velocity profiles were deconvolved into a strong component and a weak component with somewhat different properties. Figure 12.6 shows the derived velocity field of the main component, which is a nice example of the classical form for a tilted rotating disk.

The neutral hydrogen velocity distribution shows a broad peak centered at about 12 arcmin, with maximum value of 12 km/s. It decreases outward from the peak, reaching about 8 km/s at the outermost detected signals at 30 arcmin. The two-dimensional map of the main component's velocity dispersion is shown in Fig. 12.7.

These new observations revealed two second order details of the HI distribution: a suggestion of the presence of spiral density waves and clear evidence for a warp of the HI disk, similar to what had been found for several spiral galaxies.

An important breakthrough was the discovery of dark matter in galaxies. Rubin et al. (1980) showed that the rotation curves of most galaxies did not have a turndown at large distances, but remained roughly level; that is, the velocity remained constant. The explanation of this surprising observation was that some kind of invisible

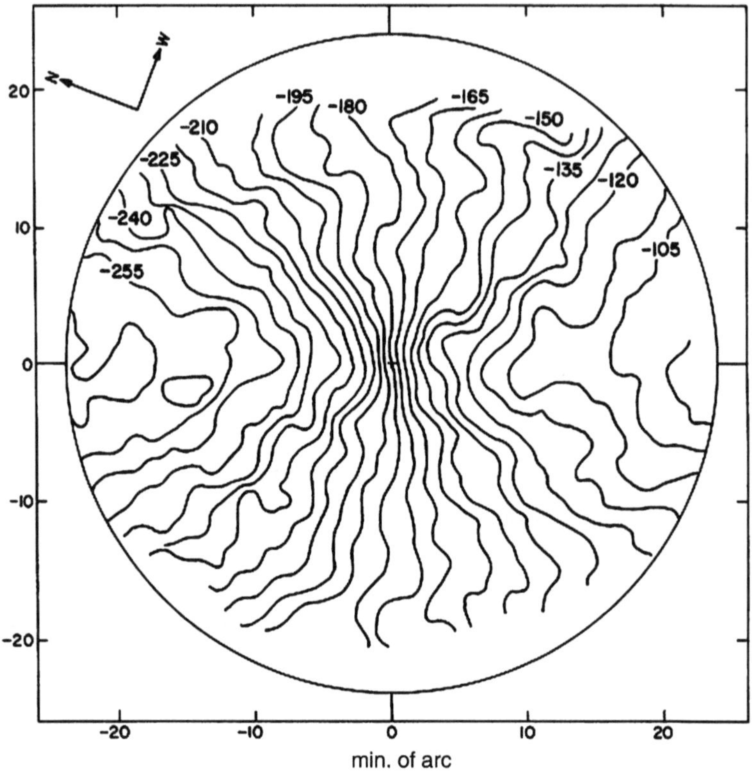

Fig. 12.6 The velocity field for the main component of the HI in M33. Velocity intervals are 15 km/s (From Rogstad et al. 1976). Used by permission, copyright AAS

mass continued to exist in the galaxy far beyond the outer visible disk (other equally exotic ideas were also proposed, including alterations of Newton's law of gravity). Much effort has been expended in the attempt to identify the nature of this invisible matter, with incomplete success, at best. It remains one of the most puzzling of science's unsolved problems. Though we do not yet know what the dark matter is, careful analyses of the velocity fields of spiral galaxies have revealed the shape of its distribution in and around the host galaxies (see, for example, Persic et al. 1996).

In the case of M33, the rotation curves determined in the early twenty-first century shows no downturn at large radii. Apparently the previous data coincidentally agreed with (or possibly were subconsciously prejudiced by) the expectation that the curve must decrease beyond the visible limits of the galaxy. Corbelli and Salucci (2000) showed that the rotation curve increases out to a distance of 18 kpc. Their best fit model implies that the dark matter dominates the dynamics from about 3 kpc and beyond and that the total mass of the dark matter is 50 billion solar masses. Figure 12.8 shows the observed HI rotation curve and the model results of Corbelli

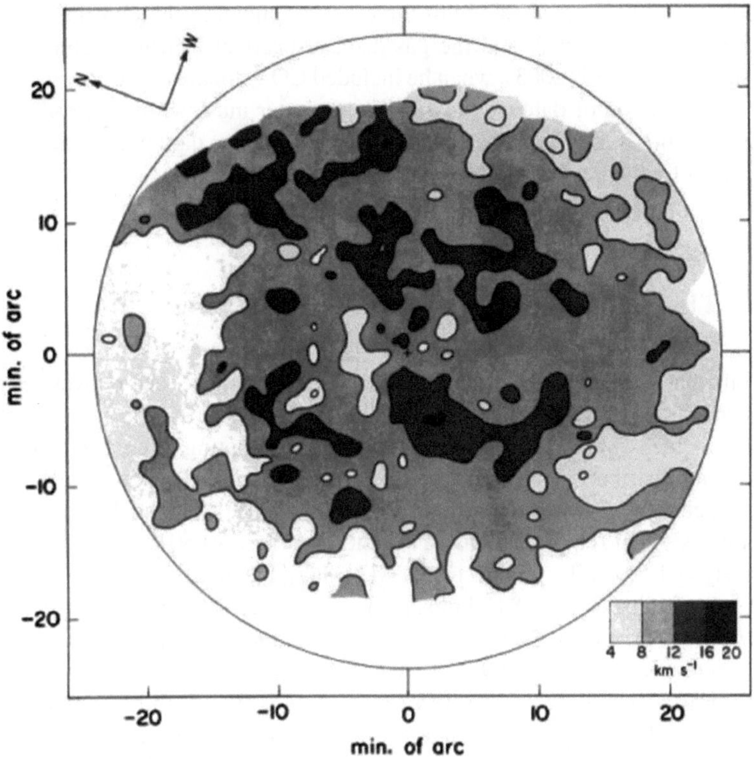

Fig. 12.7 The velocity dispersion map for the main component (Rogstad et al. 1976). Used by permission, copyright AAS

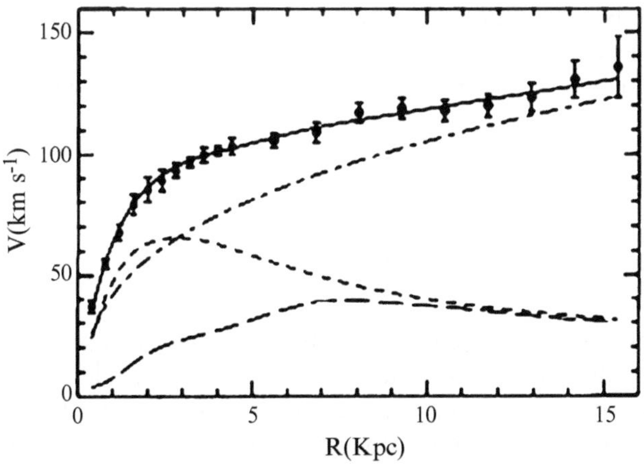

Fig. 12.8 The deconvolution of the observed rotation curve for M33 into the various components' contributions (Corbelli and Salucci 2000). The *upper curve*, which includes the observations, represents the sum of the contributions of the dark matter halo (*dash-dots*), the star component (*short dashes*) and the gas component (*long dashes*). Used by permission, copyright RAS

and Salucci's (2000) analysis, with three curves showing the contribution of the dark halo, the stellar disk and the gas disk, respectively. Further refinement was published by Corbelli (2003), when he included CO velocities. The conclusion from this new collection of data was that the dark matter mass within the radius of the disk gas is 5 times the visible baryonic mass, while a huge dark halo is indicated, the total mass of which is 50 times the barionic mass.

12.3 The Stellar Velocity Curve

The kinematics of the stellar components of M33 have been investigated by McConnachie et al. (2006). Using the Keck 10-m telescopes, the authors obtained radial velocities of hundreds of red giants. After eliminating foreground red dwarfs, the velocity distribution showed a broad spread around the systemic velocity, with some structure, which was interpreted as being evidence for three distinct kinematic components. A halo component had a velocity dispersion of 50 km/s and a metallicity of [Fe/H] = −1.5, similar to that of the M31 halo metallicity. The disk component had a velocity dispersion of 16 km/s, somewhat larger than the value of 9 km/s for HI (Warner et al. 1973). The disk stars were found to be on the average somewhat metal poor, with [Fe/H] = –0.9. The stellar disk was interpreted to be offset by a warp from the HI disk, with a difference of 15 km/s. A third component was suggested, with an offset of 50 km/s, but its character was uncertain.

The Keck Telescope giant survey was continued by Hood et al. (2008), who reported on spectra of approximately 1,000 red giants in and toward M33. Preliminary results, based on 234 probable M33 members, are that the data reveal three kinematic components: a thin disk with a velocity dispersion of 15 km/s, a thick disk with a velocity dispersion somewhat greater than the mean of the observations, 47 km/s, and a halo.

The percentage by number of stars in these components is 66%, 30% and 4%, respectively.

12.4 The Cluster Kinematics

Early studies of M33's star clusters involved spectroscopy that allowed velocities to be obtained for 45 clusters of various ages (Schommer et al. 1991). The sample divided itself into three components, similar to those described for field stars above. The old and metal-poor halo clusters have a large velocity dispersion and essentially no rotation. The thin disk cluster component has a small dispersion and rotates like the HI gas, and there is an intermediate population of clusters with a slightly higher velocity dispersion that includes objects of intermediate age.

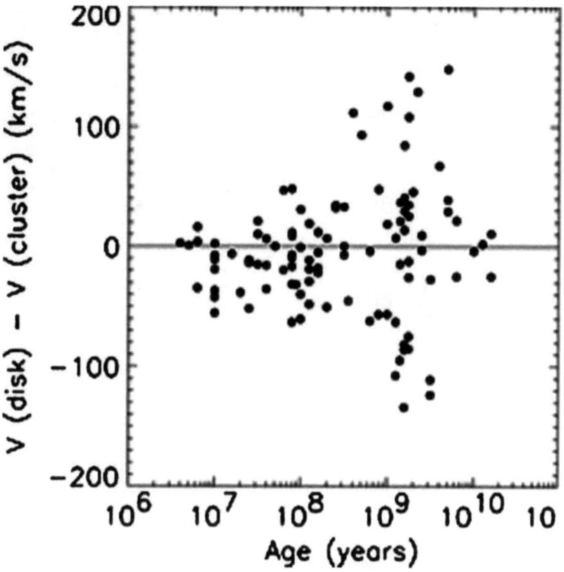

Fig. 12.9 The radial velocities as a function of age of the M33 clusters analyzed by Chandar et al. (2002). Used by permission, copyright AAS

A more complete analysis of the cluster system's dynamics resulted from a survey of velocities made with the WIYN Telescope and a multi-object spectrograph (Chandar et al. 2002). A total of 107 clusters of known age (from colors) was chosen, representing a spread in ages from very young to several Gyr old. When comparing the velocities with the ages of the clusters it was found, as had been indicated before by Schommer et al. (1991), that the young clusters showed a small dispersion of only tens of km/s, while the oldest clusters had a range of velocities nearly 300 km/s wide (Fig. 12.9). The distribution indicates that most of the old clusters belong to a halo with isotropic velocities, while 15% (±5%) of them are part of the bulge or disk population.

The young clusters in Chandar's sample were found to show a velocity curve that agrees with that obtained for other disk components, in particular, HII regions and HI (Fig. 12.10). The cluster rotational velocities are offset in velocity from the HII regions by about 15 km/s (except in the outermost part), which may indicate that the sample includes some bulge clusters. However, the errors are about this large, so this conclusion must be tentative.

To look for any relationship between kinematic population and metallicity, Chandar et al. (2002) determined approximate abundances from the spectra. They found that the old purported halo clusters showed an age spread of approximately 2 Gyr (from ~5 to 7 Gyr), which is somewhat larger than the age spread in our Galaxy and the average age is somewhat younger.

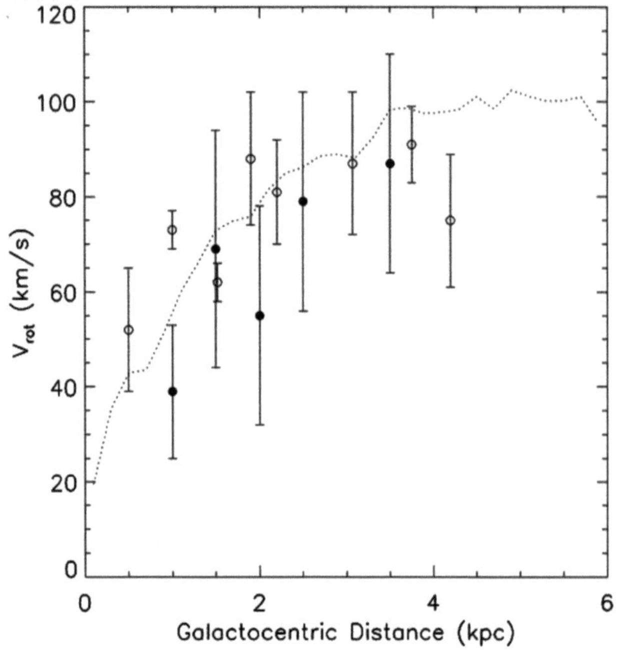

Fig. 12.10 The velocity curve for young clusters (*closed circles*), HII regions (*open circles*) and HI (*dashed line*) (The sources are Chandar et al. (2002), Zaritsky et al. (1989) and Warner et al. (1973). From Chandar et al. 2002). Used by permission, copyright AAS

12.5 Summing Up

The development of our understanding of the dynamical and kinematical character of the various components of M33 has followed a fairly straight path from the first basic steps in 1942 to the present more complicated landscape of 2011. The only jog in the path occurred in about 1980, when it was discovered that the rotation curve could not be described by any conceivable mass distribution that was confined to the distribution of the visible matter. As in most spiral galaxies, it was found that an invisible component, "dark matter", dominated the dynamics of the galaxy, especially in its outer parts. With regard to the visible matter, M33's kinematics includes three components: a non-rotating halo of old, mostly moderately metal-poor clusters, an intermediate component (possibly a thick disk) with a velocity dispersion of a few tens of km/s and a rotating thin disk of mainly young stars and gas with a small velocity dispersion.

References

Brandt, J.C.: MNRAS **129**, 309 (1965)
Burbidge, E.M., Burbidge, G.: ApJ **132**, 640 (1960)
Carranza, G., et al.: Ann.d'Astrophys **31**, 63 (1973)

Chandar, R., et al.: ApJ **564**, 712 (2002)
Corbelli, E.: MNRAS **342**, 199 (2003)
Corbelli, E., Salucci, P.: MNRAS **311**, 441 (2000)
Dieter, N.H.: AJ **67**, 222 (1962)
Hood, M., et al.: ASP Conf. Ser **374**, 281 (2008)
Lohmann, W.: Ap&SS **29**, 61 (1974)
Mayall, N.U., Aller, L.H.: ApJ **95**, 5 (1942)
McConnachie, A.W., et al.: ApJ **647**, 25 (2006)
Persic, M., Salucci, P., Stel, F.: MNRAS **281**, 27 (1996)
Rogstad, D.H., Wright, M.C.H., Lockhart, I.A.: ApJ **204**, 703 (1976)
Rubin, V.C., Ford Jr., W.K., Thonnard, N.: ApJ **238**, 471 (1980)
Schommer, R.A., et al.: AJ **101**, 873 (1991)
Volders, L.M.J.S.: Bull. Astron. Netherlands **14**, 323 (1959)
Warner, P.J., Wright, M.C.H., Baldwin, J.E.: MNRAS **163**, 163 (1973)
Wright, M.C.H., Warner, P.J., Baldwin, J.E.: MNRAS **155**, 337 (1972)
Wyse, A.B., Mayall, N.U.: ApJ **9**, 24 (1942)
Zarisky, D., Elston, R., Hill, J.M.: AJ **97**, 97 (1989)

Index

P. Hodge, *The Spiral Galaxy M33*, Astrophysics and Space Science Library 379,
DOI 10.1007/978-94-007-2025-1, © Springer Science+Business Media B.V. 2012

W

X

Y

Z

MIX
Papier aus verantwortungsvollen Quellen
Paper from responsible sources
FSC® C105338
FSC
www.fsc.org

If you have any concerns about our products,
you can contact us on
ProductSafety@springernature.com

In case Publisher is established outside the EU,
the EU authorized representative is:
Springer Nature Customer Service Center GmbH
Europaplatz 3, 69115 Heidelberg, Germany

Printed by Libri Plureos GmbH
in Hamburg, Germany